"... The purpose of this small volume is to present some information and ideas, new and old, bearing on the position of mankind in the universe of physics and sensation. ..."

"... For the most part I shall here present only the raw materials and their more obvious implications; they are to be used by the reader as he will for his orientation in the material world, and possibly in the nonmaterial. ..."

"... [We shall] adopt from the beginning the attitude of mature inquirers, and confront the cosmic facts squarely and fully: small but magnificent man face to face with enormous and magnificent universe."

—Excerpted from the Introduction

OF STARS AND MEN was originally published at $3.50 by Beacon Press.

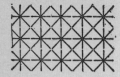

HARLOW SHAPLEY

OF STARS AND MEN

The Human Response to
An Expanding Universe

WASHINGTON SQUARE PRESS, INC.
NEW YORK

OF STARS AND MEN

Beacon Press edition published May, 1958

Washington Square Press edition published February, 1960
1st printing....................December, 1959

This Washington Square Press edition includes every word contained in the original, higher-priced edition. It is printed from
brand-new plates made from completely reset, clear, easy-to-read type.

L

Published by
Washington Square Press, Inc.: Executive Offices, 630 Fifth Avenue;
University Press Division, 32 Washington Place, New York, N.Y.

Washington Square Press editions are distributed in the U.S. by
Affiliated Publishers, Inc., 630 Fifth Avenue, New York 20, N.Y.

Contents

7. The Fourth Adjustment 95

8. A Digression on Great Moments 105

9. Toward the Emergence of Organisms 109

10. What Should Be the Human Response? 126

FOREWORD

The present treatise is not written for astronomical or other scientific technicians. It is therefore presented with few bibliographic references, and except for parts of Chapters 5, 6, and 9 it is essentially free of technicalities. When I use the phrase "Cosmic Facts," the reader is asked not to assume too rigid a meaning for the word "facts"; what is considered factual today is tomorrow recognized as capable of further refinement. The occasional repetitions of argument and phrasing in the later chapters are not oversights; they are meant to emphasize points where emphasis seems advisable.

I should record my thanks to Mrs. Shapley, who has patiently discussed details of thought and diction and assisted with the preparation of the manuscript.

If this book were to be dedicated to its first and chief encouragement it should probably salute starlight, insects, the galaxies, and the fossil plants and animals, since they have joined in inciting the analysis and attitudes herein recorded.

—HARLOW SHAPLEY

OF STARS AND MEN

$\star$ **1** $\star$

Introduction—Nothing Merely Human

~~~~~~

An attempt to build from rough materials a rational and stable structure that contains man and his universe is the goal of this writing and of the underlying studies. A considerable amount of old construction has been pulled down in recent years. Further demolition is advisable. In particular we must continue to jettison much human vanity and many anthropocentric arguments. And we should expect that our reconstruction also may need renovating in the not very distant future. Of necessity we must work in our times and for our times. We must use the concepts now available and those that can be logically developed.

Stated otherwise, the purpose of this small volume is to present some information and ideas, new and old, bearing on the position of mankind in the universe of physics and sensation. It is an essay on orientation, including a tentative obituary, one might say, of anthropocentrism in our description of the universe.

A survey of the relevant knowledge that the inquiring mind of man has brought to light naturally leads to speculative meditation, to a dreaming built out of wonderment, to a groping for both an objective and an introspective philosophy. But for the most part I shall here present only the raw materials and their more obvious implications; they are to be used by the reader as he will for his orientation in the material world, and possibly in the nonmaterial. It will be a cooperative project—writer and reader, reporter and critic.

For example, I shall emphasize the continuity in the series of living organisms—the long continuous series from the inorganic molecules, through the complex organic molecular aggregates, to the primitive plants and animals; then on through the myriads of more complex biological forms to the higher types of man—an evolution through a billion years and more. This continuity I shall stress; readers may suggest, if they will, where "spirit" appears to enter the stream of life. I shall present the expanding universe of galaxies; others may suggest what started it off, and why.

We are undertaking, in this essay on star facts and human destiny, a serious enterprise. Destiny has an ominous sound, and facts are frequently disturbing and oppressive. To postpone a possible depression about our role or function in the stellar universe, and to evade for a time the heavier implications of man's physical position, we could first emphasize the good features of life under the present constellations. We might introduce this essay in an optimistic vein, since later there will be time and need for a more somber look.

It is a good world for many of us. Nature is reasonably benign, and good will is a common human trait. There is widespread beauty, pleasing symmetry, collaboration, lawfulness, progress—all of them qualities that appeal to man-the-thinker if not always to man-the-animal. When not oppressed by hunger or cold or man-made indignities, we are inclined to contentment, sometimes to lightheartedness.

But rather than a lighthearted and somewhat evasive view of our situation and responsibilities, it would be more in keeping with what lies ahead to adopt from the beginning the attitude of mature inquirers, and confront the cosmic facts squarely and fully: small but magnificent man face to face with enormous and magnificent universe.

The first step is to ask a preliminary question and provide a synoptic answer. The question is: What is the Cosmos?

Four or More Basic Entities

The scientists who busy themselves with thoughts and researches on Cosmography, and also some philosophers whose coverage includes Cosmology, are soon led to the conclusion that the physical universe consists of, or is amply described by, or is presented to our knowledge in, four recognizable basic entities—entities that can be named and to some extent isolated. There may be more than four. But, for the sake of simplicity, we are tempted to put all the world of physics and perhaps all the biological world into the framework of these four properties. They are, of course, Space, Time, Matter, and Energy. Many quasi-entities are recognized, such as motion, velocity, metabolism, entropy, creation —but they are derivatives, or combinations among the four.

Although not yet recognized or isolated, may there not be other entities, perhaps some of even superior importance? In particular, is there one other property of the material world that is essential to make the universe go? Something like Motion in an abstract sense? To put the question in personal terms: If you were given the four basic entities and full power, opportunity, and desire, could you construct a universe like this one out of space, time, matter, and energy? Or would you require a fifth entity—another basic property or action?

We seem to belabor this point, but the somewhat mystical fifth entity must be referred to more than once in what follows. That it exists, we can hardly doubt. Is it a master entity, perhaps more basic than space and matter and possibly including them—something quite unlike the four named above?* Is it indispensable? Something that would make click a universe of stars, organisms, and natural laws that might otherwise be clickless?

Some readers may be thinking of the word and concept God, but we should not be hasty in such a deep and critical matter. Let us not use up that important and comprehensive concept for only a part of the universe, or for something already comprehensible to primitive us. Working in the field of Cosmography† we are tantalized by the possibility that hidden world characteristics do exist for which there may be an independent world-dynamic, one that we could call Direction, Form, Drive, the Will of Omnipotence, or "Consciousness." But if it be Drive or Consciousness, the concept must be of cosmic extent; nothing merely human, nothing that is only terrestrial, has a rightful place in Cosmography.

What? How? Why?

Already in the preceding pages we have revealed in several places our limited knowledge of the world. It is amazing how little we know compared with what we conceive as ultimately within our reach. And how small, probably, is that reachable portion compared with what is beyond! Superstitions and loyalty to dogmas have kept us bogged down not far above the most primitive views of the universe. Fancy how far we might now have gone if we had not been shackled by mythology and by certain social conventions and national policies; what intellectual progress we might have made by this time if we had, for example, emphasized the psyche rather

---

* We can, if we choose, combine the four in pairs: space-time and matter-energy, and search for a third basic entity, rather than a fifth.

† In rough definition, Cosmography is to the Cosmos what geography is to the earth.

than property. Eventually we may not be as poorly equipped as now to answer three basic questions:

What is the universe?

How does it work?

Why is the universe?

To the first—What?—we can respond in an elementary way, and produce a brave partial answer, mumbling something about matter, gravitation, time, protoplasm. To the second question—How?—we venture something about the laws of nature, the death of heat, the running away of the galaxies. But to the question "Why is the universe?" we are likely to exclaim "God only knows!"—and apparently it is "restricted" information. Being somewhat incorrigible, we begin to ask ourselves why the "Why?" of the universe is restricted, and why we should accept the suggestion that the now distant mysteries must remain forever in the secret category. Many of the inexplicables that afflicted the ancients are with us commonplace rational facts or operations. The element of mystery has evaporated. If we continue to work hard, to think clearly and deeply, will not some of the major why's of the universe become answerable by us? Philosophers would probably say "No!" but I let the question stand.

Plan of the Discussion

To present the principal arguments in an essay not too long, I propose to restrict the discussion to a few steps in the adjusting of our minds and acts to what I like to call the Cosmic Facts. (Perhaps I should use the word "indications" rather than the positive-sounding word "facts.") The steps we shall take are the following:

(1) An examination of the need for a new consideration of man's place and function in the Cosmos. Are there good reasons why we should now disturb our meager peace of

mind because of what science is trying to tell us? The answer is affirmative, decidedly so. We should face the revised specifications of our job, the new indications of our potentialities.

(2) The locating of the earth (and mankind) in the physical universe that we deduce from sensory data and reasoning. The search is built around the basic entities; we search out our place not only in space, but in time, matter, and energy. Is man in a commanding position, we ask, or in the ranks, or tagging along, forgotten of the stars from which he sprang? We shall show that his physical location is now definitely determined and that in a rather peculiar sense he is not wholly insignificant.

(3) The problems associated with the origin of the earth and with the number and distribution of life-bearing planets.

(4) The question of the nature and distribution of sentient life. Is it a local operation, developed under special circumstances on or near the surface of this one planet, or is it a widespread activity? (The details of life's origin and meaning are not wholly solved, to be sure, but we recognize the existence of even deeper problems.)

(5) The Fourth Adjustment in the evolution of the cosmic concepts of earth-bound man. From his primeval egocentrism in a hostile environment he has developed a view of the universe that is more in keeping with the astronomical and biological evidence. The Hostility of Nature is no longer a useful concept.

(6) A consideration of the Generic Mind, and a preliminary exploration into the sense organs as tools of comprehension, suitable for cosmographic investigation.

(7) Finally, some suggestions are offered about the future of man as a dominant terrestrial animal. This is a fertile field for guessing, also for hope and despair, for satisfaction and doubt. We are led to thoughts on the survival of the species,

and also to a consideration of human programs and the philosophical attitudes thereupon based.

In summary, this first chapter has presented the intent and plan of a discussion of man's orientation in the world as it is revealed by current advances in knowledge, chiefly in scientific fields. We have named four basic entities of the material world and have opened the door to a consideration of the structure and meaning of life. In preparation for such an inquiry we have placed before the reader the scaffolding, or at least the rough blueprints, of a cosmic castle. In the following pages we must fill in some details. We shall arrive at no completely finished structure. I am not wholly certain whether we are working on the foundation, or on the roof. More likely, and this is the modest ambition, we are working on an effective toolhouse that must precede the ultimate structure.

How might we name the edifice with which we concern ourselves? It is not science in the standard sense. Is it metaphysics, religion, human grasping, resignation, hope? Provisionally we might call it "Adjustment through Understanding."

★ 2 ★

New Bottles for New Wines

~~~~~~

We start with the following proposition: Is there not at this time a justification for a revised look at mankind as a world factor? Our prompt answer is "Yes!" To the scientist, rich in new knowledge, and to the puzzled layman, and perhaps to some philosophers, the answer is decidedly affirmative. An elementary reason for a reconsideration lies in the recognition in recent years of the "displacement" of the sun, earth, and other planets from a central place, or even a significant place, in the sidereal universe—in the placing of the observer in a very undistinguished location in a faint spiral arm of an ordinary galaxy.

This reason is elementary but momentous, for it concerns the replacement of the earlier *geocentric* and *heliocentric* theories of the universe by the *eccentric* arrangement that now we all accept. By this move we have made a long forward step in cosmic adjustment—a step that is unquestionably irreversible. We must get used to the fact that we are peripheral, that we move along with our star, the sun, in the outer part of a galaxy that is one among billions of star-rich galaxies.

If there is some grandeur in our position in space and time, I fail to find it. Our glory must lie elsewhere. Also, should we not openly question the rather vain and tedious dogma that man somehow is something very special, something superior? He may be. I hope he is. But certainly it is not in his location in space, or in his times; not in his energy content or chemical composition. He is not at all outstanding in the four basic material entities—space, time, matter, energy. Nothing unique and worthy of boast in his size, activity, composition, or his epoch in cosmic chronology. He is of course an intricate and interesting phenomenon, but we should not get sentimental about him or introspective until later chapters, and then with restraint. Once we have cleared away our illusions about man's worldly importance, we would be in a better position to consider the position of the human mind, estimate its power, its significance, and its effectiveness in grasping cosmic operations.

Egocentrism and anthropomorphic bias have long dominated our thoughts and clouded our deductions. Perhaps this is inevitable. We are human and cannot be purely objective. We have to know the world through our own sense organs. The cosmic outlook of beetles, or protons, or comets is not easy for us to conjure up and wisely exploit. We must admit, however, that the subjective approach—the thinking and acting always in terms of self or of mankind—has grave limitations. It is full of traps. Objectivity is the only brush with which to paint a true and satisfying picture of the cosmos and a clear and clean sketch of its relationship to the experiment in humanity.

The presumed superiority of man as an animal, the assumption of the importance of life, especially of human life, to the universe at large, and the insistence or feeling that our moment in the geological ages is somehow enormously significant in the flow of time—all these easy postulates should be questioned. To attain a proper evaluation of interpreter measured against the interpreted, and to counter somewhat our automatic egocentrism, we may need to overemphasize the role of galaxies and stars, those "cold fires, yet with power to burn and brand his nothingness into man."

There should be, however, nothing very humiliating about

our material inconsequentiality. Are we debased by the greater speed of the sparrow, the larger size of the hippopotamus, the keener hearing of the dog, the finer detectors of odor possessed by insects? We can easily get adjusted to all of these evidences of our inferiority and maintain a feeling of importance and well-being. We should also take the stars in our stride. We should adjust ourselves to the cosmic facts. It is a magnificent universe in which to play a part, however humble.

Many distinguished human voices of the past have spoken dispassionately of delicate man defending himself against the cold of the universe. A few have ended in puzzled despair. Some have come to the conviction that thinking man has been miscast in the cosmic drama, for he has so much to work with, to think with, yet finds himself hopelessly earth-bound, short-lived, and enslaved both by animal habits inherited from the primeval jungle and by dogmas acquired from his immediate ancestors.

Although some inquirers have kept wide-eyed, and remained hopeful that the limited mind of man can with increasing success cope with the problems of the universe, many of the meditators, perhaps most of them, soon or late, have retreated agreeably to the solace of dogmatic theology.

Provided as we are with new and basic data on the nature of the universe, we note many reasons why moderns could and should now interpret the world in a way more circumstantial and rational than it could be interpreted by Moses, or Lucretius, or Spinoza, or Locke, or Pascal, whose cosmologies were geocentric or heliocentric and limited. We have deep knowledge and much critical information that was not available to the philosophers of earlier centuries. We have gone far, very far, in the accumulation of verifiable facts. And this we must remember: there will be, if we remain civilized, no return. We must henceforth live with our scientific acquisitions. No amount of skepticism about details, no sweeping denials of observation, no distortion of the recent revelations of science can erase the intellectual progress. Wishing will not revive the dear, dead hypotheses.

A Brief Digression on Arithmetic

The Macrocosmos is so large and the elementary particles in the Microcosmos so small that the comparisons of dimensions, such as appear in the table of material organization on page 25, are cumbersome. To write down numerically our estimates of the number of fundamental particles in the space-time universe would require eighty or ninety figures—a labor to write, and a numerical idea impossible to grasp. A galaxy is a million million million million times the size of an amoeba, and an amoeba is a giant compared with the electron. Similarly, we have cumbersome comparisons when dealing with the time entity. For example, the ratio of the rotation period of a spiral galaxy to the vibration period of an ammonia molecule is greater than ten million billion billion. We need a simpler mode of expression.

To escape the impressive but clumsy numbering, we may resort to the use of power arithmetic, expressing large and small numbers in terms of the powers of the number ten. It is a very simple device, as well as convenient. The second power of ten, that is 10^2, is 100; 10^3, a thousand; 10^{-3}, a thousandth; 10^6, a million; 10^{-6}, a millionth; 10^{12}, a trillion, and so on. The number of hydrogen atoms in a gram of hydrogen is 6×10^{23}, which is six followed by 23 zeroes. The approximate age of the earth's crust is 5×10^9 years—that is, five thousand million years; 3×10^{10} is the velocity of light in centimeters per second; 10^{-29} grams per cubic centimeter is a value for the average density of matter in metagalactic space.

This number, 10^{-29}, is obviously a much neater expression than 1 over 1,000,000 followed by 23 more zeroes.

To multiply such large numbers we simply *add* the exponents: $10^{14} \times 10^{12} = 10^{26}$; if there are coefficients they are simply *multiplied*: $2 \times 10^7 \times 3.1 \times 10^7 = 6.2 \times 10^{14}$, which is the number of seconds in twenty million years.

To divide we subtract exponents: $10^{16} \div 10^2 = 10^{14}$. Additions are obvious, though infrequently used: $2.4 \times 10^7 + 1.5 \times 10^6 = 2.55 \times 10^7$.

Preview of the Argument

The advance of knowledge in the scientific realm (science widely defined) and our currently greater freedom from theological dogmas have opened up channels of analysis and avenues of thought which, if then known, would have altered profoundly the theories of the cosmogonists of earlier years and certainly would have affected their considerations of original causes. The new discoveries about life, matter, and space should be and generally are recognized as relevant to philosophy. They can be extensively documented and convincingly presented. But here it should suffice to point without much elaboration to several of the direction-bending developments. They might be called ego-shrinking revelations. Most of them will be detailed in subsequent chapters.

(1) Naturally I start with the stars and note that the number of suns now within our scientific knowledge is not the five or six thousand naked-eye objects of the ancient Hindus and Greeks, nor the million stars revealed by the small telescopes in the days of Galileo and Newton, nor the few billions of a generation ago. The number of stars in today's surveys is more than 10^{20}—a hundred thousand million billion—and each star radiates the fuel for life to whatever planets go along in the journey through the depths of space

and the eons of time. In the orientation of man and his works in the material universe the implication of that vast number of stars is obvious.

(2) No longer is the origin of life a deep mystery. Supernatural "intervention" in the biochemical development which we call life is not required. Natural operations, most of them already known, will suffice. We have bridged, at least provisionally, the gap between life and the lifeless. The microbiologist probing down from cells toward the inanimate and the chemist moving up from atoms toward the animate are practically in contact. Much detailed work, however, remains to be done. The approaches to the bridge need careful building, the structure needs strengthening, the flooring filled in. The leading workmen on the transformation of the lifeless to the living, i.e., the chemist and the biologist, have assistance from the geologist analyzing fossiliferous rocks; from the astronomer, who finds evidence for a long-enduring pre-Cambrian age; from the statistician, who shows that even the very improbable may happen, including such improbables as the "accidental" synthesis of amino acids when times are long and materials abundant. New procedures, new facts, new conclusions are here involved. In Chapter 9 this situation is further explored.

(3) Knowledge of the brevity of our Psychozoic Era* in the evolutionary history of terrestrial biology and awareness of the peripheral situation of the earth in our galaxy have now, or soon will have, one very useful consequence, namely, that thinking man's egotism is, or soon will be, under control so satisfactorily that he can look at the whole of creation more objectively than heretofore. He has no need or right to remain only earth-minded.

(4) The probably great abundance throughout the universe of highly developed forms of life, including nerve-guided

---

* Equivalent to the Pleistocene period, an interval later considered in its relation to man's evolution.

beings, indicates that we must accept ourselves and our fellow biota as only one assemblage of the biological developments, and in all likelihood not the highest development, whatever "highest" means. Because life, we shall later argue, inevitably emerges and evolves wherever the chemistry, geology, and climatology are right.

(5) The high probability of the existence of senses, and of sense organs, now unknown to man is proposed. Their existence is indeed so reasonable as to seem axiomatic. Their importance to the imagination is obvious, now that we are partially liberated from our superiority complex. Many realities may lie beyond the comprehension of human terrestrials, simply because our outfitting with sense organs is limited. This concept must be looked into seriously.

(6) The opening to exploitation of the electromagnetic radiation spectrum, which gives us an energy spread of not just one octave (violet to red) but of more than fifty octaves, has widened our knowledge of the minutiae of the atomic underworld and emphasized the richness and cosmic significance of the unseeable.

The foregoing preview of six aids to orientation are all recent outcomes of man's persistent exploration of the universe. Although some were foreshadowed by earlier discoveries, and some have always been vaguely felt, all have approached full development since the thinking, writing, and theorizing of the ancient and recent church fathers and since the building of most of the formal philosophical systems.

In summary, the discovery of the vastness of the universe that is open for life, the growing conviction that appeal to the supernatural is unnecessary for the beginning and evolution of living organisms, and the fresh ideas now evolving from other high human enterprise should be sufficient justification for a reconsideration, from time to time, of man's situation and function in the cosmos.

★ 3 ★

On Being Incidental

~~~~~~

In view of the foregoing argument, we accept the appropriateness of a further inquiry into the human response to the facts and the viewpoints revealed by current scientific research. As an initial step in the approach to the central questions about the universe—that is, to the questions "What, How, and Why"—we shall consider briefly the formal subject of Cosmography. Among other intentions, Cosmography as a research attempts to solve the most intriguing placement problem in the world—the question of the location of man in the universe of space, atoms, and light. Actually the end product of our efforts may be only an approach to knowledge of man's orientation in a complex cosmos, not an arrival. Questions without answers will be a recurrent by-product.

Again we define Cosmography loosely as the field of study that has the same relation to the cosmos as geography has to the earth.* Such a definition requires a prior definition of the cosmos, and that is difficult. We shall see later that cosmos means something more than the physical universe. Nevertheless, even though not sharply defined, Cosmography remains a science—a science with decorations. If at times it sounds a

* Cosmogony and cosmology are related words frequently confused with Cosmography, and apparently ambiguous even to the lexicographers; the first, however, generally implies pretensions to knowledge of first origins; the second is commonly defined as a branch of metaphysics.

bit like scientific philosophy, or even like a phase of religious teaching, so much the better. It will be no loss for religion and philosophy if they are infiltrated with atoms, stars, and the groping* of protozoa.

For the time being at least we shall try to keep our explorations of cosmic content and activity on the descriptive level. Although Cosmography as here presented is an elementary science, it carries a considerable intellectual voltage, enough to charge to full capacity the more sophisticated inquirer, enough to shock the casual and uninitiated.

Whatever else of significance we may later fabricate for life, it early becomes evident that the study of living things can contribute richly to Cosmography. An outstanding example is the direct association of chlorophyll with the age and structure of the sun and stars. This strange association ties the complicated chemical operation of photosynthesis with the internal anatomy of stellar bodies. The primitive plants of the Archeozoic Era, the green algae, were operating the photosynthetic apparatus more than a thousand million years ago; and the complex leaves of the late Carboniferous plants also testify to a sun power that has been essentially constant from then to now. The Paleozoic leaves testify that three hundred million years ago the solar radiation was little if any different from that we now know. The unhurried evolution of stars (at least of one star, the sun) is thus revealed by the Carboniferous ferns. A slow evolution is indicated, but how is it managed? What can be the source of the solar power that radiates energy into space at the rate of more than four million tons a second and yet does not exhaust itself over the millions of years?

The full story is too long for this essay. We simply report

---

* That word "groping" will bear watching. Protozoa are not the only animate gropers!

that to energize the ancient algae and the tree ferns of the Paleozoic, as well as modern plants, and activate the animals (including us) that are parasites on the plants, the sun transmutes hydrogen into helium and radiation, thus providing abundant energy. Fortunately for us, the radiant energy is issued by a self-regulating power plant.

The collaboration of the various sciences is here nicely shown. Geochemistry, radiology, stratigraphy, atomic physics, and astronomy combine in the clear indication that matter can dissolve into radiation. The fossil plants (and animals), we learn by the way of paleontology, indicate the constancy of the sun's heat and thus, by way of mathematical physics and astrophysics, reveal much about the internal structure of stars.

There are many other tie-ups with biology in the study of the inanimate universe. In the running of ants we can measure an energy flow that is as closely controlled by temperature as the outpouring of energy from distant stars. To study adequately the early climates of this planet we must bring together the methods and facts from a dozen scientific fields, some of them biological, some physical. When we see that many rules of nature are the same for biological cells and for chemical molecules, and when, as later elaborated, we accept the very impressive probability of millions of planets with highly developed organisms, we must conclude that the world of life should be admitted as a part of the cosmographic program.

Cosmography, when ideally described and studied, involves an extensive and complicated content. It is too comprehensive to be handled thoroughly in brief compass. It appears to be manageable, however, if used chiefly as an instrument in human orientation. In what follows in this chapter we shall report on an attempt to survey sketchily the material universe, with principal emphasis on the basic entities, and on the extent to which the exploration of them and with them appears to pinpoint terrestrial man in the over-all scheme.

Our sense organs are definitely limited in number and power, and our experience in thinking about the cosmos has extended through only a few millennia—scarcely more than a dozen of the revolutions of the outermost planet Pluto. Too much should not be expected of us. We are tyros in the project of cosmic interpretation. Our accomplishments appear to be rather substantial when we look into the past, but have we not unrolled as yet only a fringe of one page of the total Cosmic Writ?

From where we now stand in knowledge of the world it appears that the basic entities of the material universe are, as stated in Chapter 1, the simple-sounding "qualities" or entities of *space, time, matter,* and *energy.* Of the four, we note that matter and energy are two forms of the same thing, tied together with the most popular equation of our times (after $2 + 2 = 4$), namely, $E = Mc^2$. That equation says that to transform mass, M, into energy, E, or energy into mass in a quantitative fashion we simply apply the square of that most fundamental of natural units, c, the velocity of light. By way of the relativity theory, also space and time are now commonly united as space-time. For our present practical approach, however, we shall ignore these postulates of equivalence that arrange our entities in two pairs and consider each one separately. But first, a few remarks on the simple technologies of human understanding.

The Four Elementary Alphabets

History records that the human cultures of the past few millennia have been based to a large extent on the use of some simple aids to communication. These aids we shall call *alphabets,* widening the meaning of that word, since more than the ABC's are involved. Without the alphabets we could not readily ask and answer on a high level or communicate readily from the present to our posterity. Since their invention or emergence, alphabets have enabled men to coordinate better their knowledge and ideas, and to comprehend many phases of the surrounding complex world. They have served to reduce the seeming chaos and to lay the foundation for civilized cooperation among individuals and groups. The alphabets have also enabled men to advance their cultures and build stable societies.

The primitive grunts, squeaks, and gestures that man brought up from the "jungles" did not long suffice for such a mind-evolving primate. He had no marvelous antennae with which to communicate to his fellows, such as those possessed by the ants and used by them in building up their elaborate societies. Evolving man, if he was to survive and grow as a cosmic interpreter, had to devise and use symbols for social communication, and he had to do so more effectively than did the other animals and his own jungle ancestors. For effective communication he had to associate these symbols with sounds and ideas. He needed, and many times did design, tables of symbols to aid in social collaboration. In brief, to build his colonies and eventually his intercolony cultures it became essential to devise and introduce writing, reading, and arithmetic.

A few thousand years ago the elementary alphabets began to appear. They came in the form of ABC's and the 1, 2, 3's. The letters could be formed into words to represent ideas, and in the various isolated cultures the words became standardized. They were formed into phrases, the phrases into sentences, and in some of the higher cultures the sentences were assembled into chapters, books, and libraries.

The number alphabet was basic in primitive economics,

and, with the ABC's, eventually produced the business operations of the modern world. The numbers led to our system of weights and measures. Without these alphabets—the letters and the numbers—we would culturally be little advanced beyond the birds, bees, and apes.

Two other elementary alphabets have long existed. One is connected with the entity Time and the other with the entity Space. They are, respectively, the calendars of days, weeks, months, and years, and the maps that record space measures on the face of the earth, that is, record the terrestrial latitudes and longitudes which permit the delimitation of fields, cities, and states.

These elementary alphabets no longer suffice, either in the study of Cosmography, or in any general effort of trying to understand a world that has become enormously rich in information content. They met our needs up to a century or so ago. With the growth in amount of information, however, it has become necessary to supplement the elementary alphabets, and introduce logical classifications. Well-organized, small tabular categories have been set up to facilitate the acquiring of knowledge about stars, atoms, plant varieties, rock series, and the like. These tables, in a way, are minor and specialized alphabets.

To assist further in our study, it is now proposed to construct a major comprehensive alphabet for each of the four entities: time, matter, space, and energy. Through the use of these tabulations we shall simplify the natural complexities arising from so much specific information. Fortunately, two of these basic tabulations are already at hand, perfected and in professional use. They are the periodic table of the elements for matter, and the geological age scale for time. The former concerns matter in its elemental forms; the latter, time in large chunks. We shall begin, however, with a new table that is specially designed to attain our orientation in space—but first a digression.

On the Goals of Cosmography

As scientists and dreamers we are curious about our position in the plan of the universe. Curious also about the "planning," and sometimes inclined to talk about the planner. It is a fascinating enterprise. We can have a stimulating and in the end a satisfying experience in contemplating cosmographic facts and speculating on human fate and fancy.

The orientation of man is of course an absorbing subject, in part because he is an awkward and somewhat vain animal, but more because he is, whether he knows it or not, aimed at the stars. However ruthless he may have been in his jungle childhood and during his nonsocial past, he is now instinctively ethical, not so much because virtue may please his tribal gods but because it is good economic and social policy. He is bent also on comprehension. Moreover, to make an anticipatory statement, man now knows that he is participating, at a high and complex level, in a great evolutionary drive; he is going along, for the most part cheerfully, with such companions as the vibrating atoms, the radiating stars, the condensing nebulae, the groping protozoa, and the perennial forests with their aspiring birds and butterflies.

As cosmographers we enjoy the decipherment of some of the rules of the cosmic game. We salute the biological winners when we recognize them, such as the fish and the club mosses which can trace their ancestry of unchanged forms through many geological periods; and we can try to understand the losers, such as the trilobites of the early Paleozoic, the dinosaurs of a hundred million years ago, and Neanderthal Man.

We also occasionally venture to the borders of science to seek deep answers and to discuss our hope of contributing to

future ages something more than our fragmented skulls in the
fossiliferous rock. Naturally we are proud of the varied beauty
of human thought and action, proud of our poetry and song.
We are actors in a great cosmic play where the performers include the atoms, the galaxies, and the eternal intangibles.

The prophets of ancient Israel gloried at times in the mag-
nificence of the universe, which of course, in their time, was
centered on man. Those days, however, were scientifically
very early and chronologically perhaps more than a third of
the way back to the beginning of human cultures. What the
inquiring mind has since uncovered would have been incredi-
ble if revealed to the ancient prophets. Their vision was, we
now see, myopic. Our vision is doubtless also deficient, but at
least we recognize that we are taking part in a play far grander
than foretold in ancient times. The advance notices of two or
three millennia ago greatly underestimated the cosmic drama.
Reverence then had to be supported with imaginings and
superstition. But the accepted facts of now far transcend the
fictions of not so long ago. So it seems, at any rate, to those
who look downward into atoms and the biological cell and up-
ward to the stars. To be reverent, we now have no need of
superstitial aid.

In our cosmic inquiries we may appear boastful with regard
to the inadequacies of the ancient philosophies, but we should
suffer a healthy pride-shrinking experience in foreseeing that
a century hence we, too, may be considered to have been
primitives in knowledge and thought. Indeed, two of the present goals of the exploration among galaxies and atoms are the
same goals that should prevail in other fields of science, namely, to strengthen the evidence on which we can construct our
current understanding and to contribute through research as
rapidly as possible to the obsolescence of our presently cher-
ished hypotheses. We hope for greater knowledge and sounder
ideas in the future. Deeper thoughts will surely come, wider
spread of the senses, fuller appreciation of the functioning of

the human brain, higher ambitions for men participating in the greatest operation of nature—an operation of cosmic dimensions that might simply be called Growth.

With this glimpse of the welcome challenge, we now return to the basic alphabets of Cosmography—return to what might be called the Tables of Higher Orientation.

Spatial Dimensions—An Alphabet for Organizations in Nature

We begin the consideration of organizations in nature with a look at the varied types of *human* associations, thus flattering a group that will have few firsts. The simplest organization is the family. Maintaining most of their individualities, families are naturally grouped into neighborhoods or villages. Not many of the natural freedoms are sacrificed in exchange for the advantages of this higher organization. But villages and neighborhoods rarely remain entirely independent of each other; they form states and nations, and the individual becomes less autonomous and free as the higher grouping develops. Beyond the nation we may eventually have, on this planet, a world state, or at least a world-wide cooperating civilization.

A further step in societal organization, beyond the human family and the association of families and groups, is the integrated society. It has appeared among the protozoa Volvox, at low levels of assembly and social coordination, and among the social insects (bees, wasps, ants, and termites) at the highest level attained in the animal kingdom.

Wherever we find organization we can of course assume the existence of an organizing force or cause. In the human family the binding force is love, broadly said, and for our higher group associations, from village to nation, it is security, principally, but also cultural ambitions and the desire for an abundant economic life. With the social insects the binding attraction is complicated and not fully understood, but the taste organs are doubtless involved, and among the ants the interchange of swallowed food.

In the microcosmic *inanimate* world the organizational tendencies are governed by electrostatic and similar forces. The fundamental particles (electrons, neutrons, etc.) organize first into atoms, then the atoms into molecules and molecular systems, and on to larger and more complex organizations in the crystals and colloids.

With each higher organization, as with human society, the elemental freedoms are increasingly cramped. The wild liberties of an atom in the vacuum of intergalactic space have been largely lost to the molecules of oxygen and nitrogen in a closed room, where the air is earth-bound by gravitation and the incessant collisions prevent any one molecule from getting anywhere. And the atomic units of the solid metals of my pencil are so much further cramped, confined, and controlled that I do have a pencil. It is composed of agitated electrons, protons, and neutrons that cannot freely escape into the interplanetary spaces. Always for the advantages of organization, the price is the loss of elemental freedoms. A civilization, for example, costs much in loss of individual liberties, since excesses in freedom would lead to deficits in security and social advantage. The freely roaming and unpoliced dinosaurs, I like to remember, had no plumbing.

As we proceed to greater and more massive organizations, we begin to leave the microcosmos at the colloidal level. We enter the macrocosmic world where gravitation appears as the effective control. In the sidereal realm, we find dust particles assembling into proto-stars, with gas pressure and radiation opposing the dominating gravitation. Stars frequently appear in pairs and triples. Star clusters and clouds of stars are next in magnitude of stellar organization, and these, with the scattered single and multiple stars, are congregated into the great cosmic units called galaxies. A continuous sequence, we find, from atoms to the Metagalaxy.

TABLE I

A Classification of Material Systems

—5 ......

—4 Corpuscles (Fundamental Particles)

α.	η. Mesons, 1 to x
β. Radiation quanta	θ. Neutrinos
γ. Electrons	ι. Antineutrinos?
δ. Protons	κ. Antiprotons
ε. Neutrons	λ. ......
ζ. Positrons	

—3 Atoms
    0 to 101+

—2 Molecules
    1 to n

—1 Molecular Systems
    I. Crystals
    II. Colloids

  0 Colloidal and Crystallic Aggregates
    α. Inorganic (minerals, meteorites, etc.)
    β. Organic (organisms, colonies, etc.)

+1 Meteoritic Associations
    1. Meteor Streams
    2. Comets
    3. Coherent Nebulosities

+2 Satellitic Systems
    I. Earth-Moon Type
    II. Jovian Type
    III. Saturnian Type

+3 Stars and Star Families
    α. Stars with Secondaries
        I. With Coronae, Meteors, and Comets
        II. With Nebulous Envelopes
        III. With Planets and Satellites

TABLE I (Continued)

β. Stars with Equals
    I. Close Pairs (or Multiples)
      a. Eclipsing
      b. Spectroscopic
    II. Wide Pairs (or Multiples)
      (α) Gravitational
      [(β) Optical]
    III. Motion Affiliates

+4 Stellar Clusters
    α. Open
      [a. Field Irregularities]
      b. Associations
      c. Loose Groups
      d. Compact Groups
      e. Dense Groups
    β. Globular
      I. Most Concentrated
      II. . . . . . .
      . . . .
      XII. Least Concentrated

+5 Galaxies
    A. Bright
      I. Irregular (I)
      II. Spiral (S)
        α. Abnormal (Sp)
        β. Barred (SB)
          (I) Open (SBc)
          (II) Medium (SBb)
          (III) Concentrated (SBa)
        γ. Regular (S)
          (I) Arms Very Wide (Sd)
          (II) Arms Wide (Sc)
          (III) Arms Close (Sb)
          (IV) Arms Very Close (Sa)
      III. Spheroidal (E)
        a. Most Elongated (E7)
        b. Less Elongated (E6)
        . . . .
        g. Least Elongated (E1)
        h. Circular Outline (E0)

TABLE I (Continued)

B. Faint (Bruce classification)
   Concentration and Shape
   a1 a2 a3 ...... a10
   b1 b2 b3 ...... b10
   ....
   f1 f2 f3 .... f10

+6 Galaxy Aggregations
   1. Doubles
   2. Groups
   3. Clusters
   4. Clouds
   [5. Field Irregularities]

+7 The Metagalaxy
   α. Organized Sidereal Bodies and Systems
      1. Meteors   4. Stars
      2. Satellites   5. Clusters
      3. Planets   6. Galaxies

   β. The Cosmoplasma or Matrix
      (α). Interstellar Particles
            1. Cosmic Dust and Meteors
            2. Diffused Nebulosity (dark)
      (β). Interstellar Gas
            1. Corpuscles
            2. Atoms
            3. Molecules
      (γ). Radiation
      (δ). ....

+8 The Universe: Space-Time Complex

+9 ....

Subdivision symbols:

α, β, γ.............. differences largely dependent on basic nature
0, 1, 2, 3............ differences largely dependent on size or mass
I, II, III.............. differences largely dependent on structure
A, B a, b............ differences largely dependent on position of observer

The three groups in square brackets are chance associations, not gravitational systems.

The ever higher organizing of organizations thus appears to be a basic tendency of Nature, and making use of this property we can construct an important alphabet—a table to define our own place, our own orientation in the entity Space.

The system of material organizations is displayed in Table I. Everything material is included, even the undiscovered fundamental particles and the transcendent superorganizations. The major subdivisions are shown for all classes. The entry —3, *Atoms,* appears later, subdivided in detail, as the alphabet of matter, with its hundred different species.

Such a listing of organizations, arranged in order of increasing average dimensions, provides the framework upon which to spread ideas concerning the compulsions and attractions involved when individuals, whether physical or biological, form groups.

The *sub*division of some of the numbered organizations would be extensive and instructive, for instance, the biological groupings that here appear only parenthetically in subclass β of Class Zero. In our table of cosmic organizations we can allow little space for organisms, if we are to preserve a fair balance. There is too much known for tabling all biological forms. A full display of the ancestry of man would run in decreasing inclusiveness through ten categories, each of which would require subdividing:

Terrestrial Animate Nature (plants, animals, protista)
    Kingdom: Animal
    Phylum: Chordata (one of about 15 phyla in the animal kingdom)
    Subphylum: Vertebrata
    Class: Mammalia
    Order: Primates
    Family: Hominidae (we are here drawing away from the apes)
    Genus: Homo
    Species: Sapiens
    Variety or race: e.g., Caucasian
    Individuals

And there are various intermediate categories, such as sub-class, superfamily, and so forth.

The negatively numbered groups in the table of material systems are in the part of the sequence we call the Micro-cosmos; the positively numbered are macrocosmic. Electro-static and molecular forces, as noted above, govern the organization in the Microcosmos, gravitation in the Macrocosmos. Counteracting the molecular and the gravitational attractions are the dispersing forces: radiation, gas pressure, electrostatic repulsion, and "cosmic repulsion." This last is the name we give to the somewhat vague operator that forces the super-organization of galaxies to disperse, except in regions of clustered galaxies where gravitation still weakly controls the situation.

We might question the propriety of entering −4, Fundamental Particles, in a listing of material systems. Are these corpuscles actual organizations or are they indivisible units? They are here listed as organizations in anticipation of the submicroscopic analyses in the future which may reveal the structure of electrons and protons. Neutrons, in a sense, are already recognized as composite. In any case, the fundamental particles should be included in the table so that we may have here a complete story of the known material structures.

Classes −5 and +9, now empty, are challenges to the future. There is so much curiosity and ingenuity within the parentheses of Class Zero, subclass β, that it would be unwise to close the doors, at either end of the series, to the possibility of future revelations. We also leave openings to accommodate further discovery among the basic particles, and likewise in the contents of interstellar space. As to the former, we are by this challenge asking if there may not be something more fundamental than radiation quanta, or if there may not be other fundamental particles not included in our generous listing. As to the latter, in the fourth part of the Cosmoplasma or Matrix, we are wondering if there may be something specific and measurable in interstellar space besides particles, gases, and radiation.* A generation ago we considered atoms to be made up of electrons and protons only. Now look at the listing of fundamental particles. The category mesons, with which I associate the V-particles, includes a dozen or more evanescent but nevertheless fundamental particles.

Geological Ages—A Higher Alphabet for Time

Looking again at our listing of material systems that extend from subelectrons to the space-time complex and beyond, we are impressed by the fact that motion prevails throughout the long series. Everything is moving. The movements are relative to various zeroes or coordinates, or with respect to bodies of similar or different character. The radiation quantum represents an energy transfer in space with the speed of light, and the electrons in the atom also move at tremendous speeds, according to commonly accepted atomic models. But such high velocities are not common. Some of the relative motions, like those of crystals fixed in rock, differ immeasurably from zero. Other slow motions are those of plants and animals on a planetary surface. Intermediate speeds include the velocity of a comet in passage around a star, and the recession of nearby galaxies in an expanding universe.

Notwithstanding its universality, motion is hardly a fundamental or basic entity of the material universe. It is a change of position, and the speed of a change is measured as space (length) divided by time. The prevalence of motion everywhere emphasizes, therefore, that time is a basic factor in the career of material systems. Growth and decay are time-linked. Organizations can fade away. Comets, for example, dissolve;

---

* Space quanta? Instons of time? Emergons? Psychons?

open star clusters are slowly dismembered by gravitational shearing; molecules are forced by radiation to dissociate; organic bodies rot, and nations decay. Also, organizations of all sorts, physical and biological, have emerged in the course of time out of uncoordinated prestuff of various kinds. Most of them grow slowly in complexity and volume, some speedily by mutations.

The time element everywhere enters the panorama of the universe. We can aid our understanding of origins and growth, of decay and death by an *alphabetization of time* intervals, much as our space concepts are aided by the table of material systems.

For the full description of operations in the material universe that involve temporal sequences, we need a very comprehensive calendar. In fact, we need many kinds of clocks and calendars, tuned to the many various needs. Those that now conveniently hang in the office and home are of no use in timing the laboratory's transmutation of hydrogen atoms into helium or the explosive release of the atomic energy that runs the stars. Nor on the other hand are such calendars useful in considering the relatively slow evolution of beetles or the rotation of a galaxy. The intervals are too coarse for electrons, too fine for mountain building. For the latter, however, a most impressive calendar is already at hand—the geological age table.

One of the most fortunate breaks that has come to inquiring mankind, in addition to the provision of his relatively large cerebral cortex, is his having evolved on the surface of a planet that is extremely old. Probably the first life forms, and certainly the oldest rocks of the earth's crust, were already in existence in the very early days of our expanding universe. When the trilobites dominated the shallow seas, the galaxies were much closer together than in present years. Many of our brightest stars, it is now believed, were born long after the great Mesozoic lizards disappeared. We can, if we will, use our geological calendar astronomically and speak of Pliocene stars and Cretaceous galaxies.

If our planet and its sentient life, including mankind, had been the product of a recent sidereal event, say 5,000 years ago, instead of the product of stellar violences that date back 5×10^9 years, we would have found it hard to discover the origin of stellar energy and to estimate the age of the stars. Our base line would have been too short. We are indeed fortunate to be established on a relatively steady and very ancient crust.

Since prehistoric times the rotation of the earth in the vacuum of surrounding space has been accepted as our best timekeeper. In pre-Copernican times nearly everyone misidentified the rotation as the daily trip of the sun around the fixed earth; but that presumption only transferred timekeeping to the sun and stars. The earth's rotation was and is measured against the distant stars, used as fixed reference points, and its period, the day, is known with astounding accuracy—to a millionth of a second. But that is not good enough for modern science. The earth's rotation is slightly disturbed by the variable distribution each year of polar snow and ice; and there are internal adjustments of the rock layers beneath the surface of the earth which can also affect its regularity. Moreover, the moon, through its production of tides in the earth's air, water, and land, acts as a brake on the rotation; the same for the sun, but less effective because more distant.

The earth's incompetence as a keeper of the most highly accurate time has incited the development of ingenious timekeepers, such as those involving the very precise pendulum slave-clocks, and the vibrations of crystals and of the atoms inside the molecules of ammonia. Other atomic timepieces are currently under development.

The pulsating and eclipsing variable stars are also celestial timekeepers but, in practice, of low accuracy; and likewise the circling satellites of Jupiter and Saturn. The revolving of our sun around the center of the galaxy provides a time unit of some two hundred million terrestrial years, with an uncertainty, however, not in seconds or days or months, but in the millions of years. This big time unit, the cosmic year, even though only roughly known, is of interest when we consider the transformation of galaxies from one type to another. It enters in predicting the time necessary to dissolve the Pleiades, and in speculating on the age of our Milky Way.

The most impressive and useful calendar for Cosmography, however, is that provided by the spontaneous and natural decay of uranium, thorium, and other radioactive elements that are embedded in the rocks of the earth's crust. Paradoxically, the micro-micro-seconds of the radioactive atomic transformations are involved in the construction of the geological calendar for which the millennia are the time units. We use the briefest to measure the longest.

As with the timing of a galactic rotation, the percentage accuracy for geological dating is not high. Nevertheless this Calendar of the Eras is one of the prize threads of information that man has laboriously unraveled.

Associated with the radioactive rocks, where the automatically decaying atoms of uranium grow fewer with time while the end products, helium and lead, grow more, we find the fossils of ancient life. We find bones, sometimes, and shells, leaf traces, seeds, and tracks in the ancient fossilized sands and mud. We properly assume that the age of the rocks, which radioactivity measures, is also the age of the fossils. From their own standpoint the animals and plants of past eras are very dead, but they are exceedingly alive in our reconstruction of the story and tempo of biological evolution. The distribution, nature, and age of these fossils also assist in solving the puzzles of the origin of our planet and the secrets of its early days. Again we note that some of the fossil plants in the rocks testify eloquently to the long dependability of sunlight as we know it.

TABLE II

Geological Time Table

ERA and Period	Time in 10⁶ years since beginning	Representative and Dominant Organisms
PSYCHOZOIC		
Pleistocene	1	Primates, Insects, Flowers, Fishes, Birds
CENOZOIC		
Pliocene	15	
Miocene	35	Mammals, Grasses, Birds, Insects, and Flow-
Oligocene	50	ering Plants; Turtles, Fishes, Snakes, and
Eocene	60	Crocodiles
Paleocene	70	
MESOZOIC		
Cretaceous	120	First Birds, Mammals, Flowers, and Decidu-
Jurassic	170	ous Trees; Dinosaurs, Fishes, Cycads, In-
Triassic	200	sects, Ammonites
PALEOZOIC		
Permian	220	Fishes, Ferns, Frogs, Corals, Crinoids, Early
Pennsylvanian	240	Conifers, First Insects
Mississippian	260	
Devonian	310	
Silurian	350	Algae, Corals, Starfish, Crinoids, First Ferns,
Ordovician	400	First Fishes, Clams, Snails, and Trilobites
Cambrian	500	
PROTEROZOIC	(1,000)	Algae and First Sponges
ARCHEOZOIC	(2,000)	First Algae
COSMIC	(5,000)	Preorganic

The geological time table, as it has now been worked out by investigators in geology, radiology, paleontology, geophysics, and geochemistry, provides a fairly good calendar back to the beginning of the Cambrian Era five hundred million years ago (Table II). It also roughly dates the much older igneous rocks that are associated with the most ancient dim records of simple algae and fungi. These records are scanty and not too sure, but they suggest an age of at least fifteen hundred million years for organisms that knew how to use sunlight for energy.

Although uranium, radium, thorium, helium, and lead were the principal original elements involved in the construction of the radioactivity-based cosmic calendar, several other chemical elements are now used in the dating of fossils, rocks, and human artifacts. Among them are potassium decaying into calcium and argon, rubidium into strontium, and the relative abundance of isotopes of oxygen and of carbon. Probably additional elements will become useful as techniques improve. The time table has increasing validity.

The geological eras and periods, with the times of their beginnings, and a reference to their biology, are given in Table II without subdivision and without further comment at this point on the significance of rock ages in Cosmography.

The Periodic Table of the Species of Atoms

The third of our four tables is probably the most compact and meaningful compilation of knowledge that man has yet devised. The periodic table of the chemical elements does for matter what the geological age table does for cosmic time. Its history is the story of man's great conquests in the microcosmos. Following the pioneer work of Newlands, Mayer, and especially Mendeléev, an inspired band of workers in chemistry and physics has brought to essential completeness this basic categorizing of atoms.

The tabulation presents the now known species of atoms, arranged in vertical *groups* and horizontal *series*, and, when given in full display, supplies much information about atomic structure.* The table encompasses all the kinds of matter, from the hydrogen of atomic number 1 through helium, car-

---

* The symbols are abbreviations of the names of the elements, which are listed in all modern chemistry text books.

bon, oxygen, iron, silver, gold, uranium (atomic number 92) up to the several unstable elements heavier than uranium which are creations of our atomic power houses. The last three are named for Einstein, Fermi, and Mendeléev.

TABLE III

Periodic Table of Elements

1 H																	2 He
3 Li	4 Be	5 B	6 C										7 N	8 O	9 F	10 Ne	
11 Na	12 Mg	13 Al	14 Si										15 P	16 S	17 Cl	18 An	
19 K	20 Ca	21 Sc	22 Ti	23 V	24 Cr	25 Mn	26 Fe	27 Co	28 Ni	29 Cu	30 Zn	31 Ga	32 Ge	33 As	34 Se	35 Br	36 Kr
37 Rb	38 Sr	39 Y	40 Zr	41 Nb	42 Mo	43 Tc	44 Ru	45 Rh	46 Pd	47 Ag	48 Cd	49 In	50 Sn	51 Sb	52 Te	53 I	54 Xe
55 Cs	56 Ba	57:72 La : Hf		73 Ta	74 W	75 Re	76 Os	77 Ir	78 Pt	79 Au	80 Hg	81 Tl	82 Pb	83 Bi	84 Po	85 At	86 Rn
87 Fr	88 Ra	89 Ac	90 Th	91 Pa	92 U	93 Np	94 Pu	95 Am	96 Cm	97 Bk	98 Cf	99 Es	100 Fm	101 Mv	102 No	......	......

Not only have the scientists of the past century constructed this complete two-dimensional coherent alphabet of matter but, through the production and identification of scores of *isotopes*, they have produced for it a third dimension. Thanks chiefly to the transmuting powers of the modern "atom-smashers," all the kinds of atoms can be made to appear in isotopic form, that is, with variously weighted nuclei. For example, the naturally radioactive uranium atom may weigh either 238 or 235 units. Mercury has ten isotopes, seven of them stable. Tin has ten stable and seven radioactive isotopes. Many isotopes occur naturally; still more are only man-made. Although the atomic weights differ for the various isotopes of an element, its outer structure of electrons, and therefore its chemical and spectroscopic properties, are essentially identical from one isotope to another.

For most of the elements the artificial isotopes, made in the cyclotrons, are short-lived, vanishing through radioactive decay in small fractions of a second. The dangerous hydrogen bomb by-product, strontium 90, is, alas, not an evanescent isotope; for scores of years after its explosive creation it remains a poisonous menace.

The radioactive isotopes of many common elements underlie the tracer techniques that are now so potent in medical diagnosis and therapy, and in biological research. They are also increasingly important in geology. To use an analogy, we may say that the tracer elements radium, lead, rubidium, etc., provide a diagnosis of the aging of rocks; and therefore, by way of fossil plants, provide a chart of the past vitality of the sun.

Without the principles and practical knowledge underlying the groups and series of the periodic table, the modern industrial age would not have been possible. And on the "impractical" side, no tabulation could better illustrate the value of the higher alphabets for the orientation of mankind in the material universe.

For various reasons the student of cosmic chemistry should be familiar with the bright stars and their spectra. The stars have influenced the philosophical thought of man from the earliest civilizations. They are at the beginning of man's lesson on his place in the universe. Also, they are high-temperature laboratories in which to test not only the properties of atoms but also the skill of the spectroscopic scientist.

More than sixty of the hundred kinds of atoms known on the earth's surface register also in the solar spectrum. The spectra of stars are equally revealing. Doubtless the other elements exist in the sun and stars but are not easily detected.

Many of the man-made isotopes, however, are probably terrestrial only, or, if in the sun, are not near the solar surface.

We have no evidence as yet of strange chemistries in the stellar laboratories scattered throughout space. The calcium and hydrogen atoms in the most remote of the receding galaxies appear to react as they do at the sun's surface and in the laboratories of terrestrial investigators. Even man-made technetium, number 43 of the periodic table, is now identified in the atmospheric spectra of some peculiar distant stars. Since technetium is radioactive with a relatively short life, it must be currently manufactured by some as yet undisclosed process near the stellar surface, perhaps in "star spots."

Throughout the accessible universe there appears to be a common chemical composition (though relative abundances differ from star to star); and everywhere similarity in atomic behavior prevails.

The World-wide Argon Traffic

The helium, neon, argon, krypton, and other inert gases that line up in the last column of the periodic table are minutely present in our atmosphere. They remain entirely free of entangling alliances, unlike the atoms of oxygen and nitrogen which form combinations with many elements and particularly with the carbon on which life is built. Except for argon, these so-called noble gases appear only as traces, altogether about one-thousandth of one per cent of the earth's air.

In our atmosphere the atoms of argon, on the other hand, are about half of one per cent of the whole population of atoms. They are thoroughly mixed with the oxygen and nitrogen, and become a medium of exchange between all air breathers of the past, present, and future. They neither perish nor yield their individuality to molecular combinations. They do not escape into interplanetary space as do the lighter atoms of hydrogen and helium. Their origin is as one product of the natural radioactivity of one of the isotopes of potassium.

With each breath that we or any other man-sized animal breathes, forty thousand million billion argon atoms are inhaled, and then, without loss, since they do not combine with anything, they are exhaled for rapid and thorough diffusion by the winds throughout the earth's atmosphere. Some of the argon atoms breathed in his first day by Adam (or any early man) are in the next breath of all of us. Some of the argon of our today's breathing will be in the first gasp of all infants a century hence. This argon traffic is obviously rich in suggestion; it implies a droll one-worldness and, like sunshine, recognizes no national boundaries. It links us with the breathing animals of the remote past and distant future in a sort of communal way.

The Ether Spectrum—An Alphabet for Energy

A table that simplifies the consideration of energy will complete the collection of major cosmographic aids. Such a tabulation is derivable from the so-called ether spectrum, or electromagnetic spectrum of radiation. It is not as comprehensive and satisfactory as the tables available for space, time, and matter. There are energies, gravitational and mechanical, that are not directly included in the radiation sequence. But for exploring and understanding the total universe the most revealing energies are recorded in the electromagnetic spectrum. It was radiant energy that made possible the origin of terrestrial life and its continuation. Our existence, our warmth, our food, and most of our knowledge now depend on the solar energy transmitted through a short section of the ether spectrum. (Atavistic sun worship should be natural for us.)

Leaving out the subdivisions, we can exhibit this tabulation

TABLE IV

The Radiation Sequence

Cosmic rays, Primary	(−20)	Visual light	(−6.4 to −6.2)
Secondary	(−15)	Infrared	(−6.2 to −4)
Gamma radiation	(−12)	Microwaves	(−3 to +1)
X rays, Diagnostic	(−10)	Radio	(+1 to +4)
Soft	(−8)	Power, Light	(+7)
Ultraviolet, Hard	(−7)	Macrowaves	(+8 ...)
Soft	(−6.5)		

(Representative wave lengths in meters, expressed
as powers of 10; see page 11)

in brief and simple form (Table IV). We should note, with high respect for man's intellect and industry, that he himself with artificial sense organs has extended the recognizable visual sequence far beyond that known throughout all human history until a century ago. His knowledge and use of radiant energy is no longer confined to the small violet-to-red segment. It ranges beyond the violet through ultraviolet and X radiation to gamma rays. It goes beyond the red and infrared to radio and to the macrowaves of the light and power services of home and industry. His supplementary "sense organs" that permit this extension include photographic emulsions, thermocouples, photon tubes, transistors, oscilloscopes, Geiger counters, cloud chambers, and a maze of other electronic gadgetry. Pretty good for a recently arrived primate!

The full discussion of energy as a fundamental entity would include many recent scientific developments of relevance to Cosmography. It would detail the steps taken by physicists, astronomers, and engineers in extending spectrum analysis down into the short wave high-energy radiations. It would report how man's visual organs, the eyes, have been gradually supplemented by the ingenious accessories noted above. Nature's provision has been far transcended.

Scores of hitherto untouchable octaves of the electromagnetic spectrum, to left and to right, have been explored. X-rays were discovered sixty years ago and quickly put to use to advance human health and human knowledge. The invisible ultraviolet has become a tool of industry and medicine and an aid to research into the nature of molecules and of biological cells. Equally great human service and industrial development has resulted from the extension of that original one octave of visual light into the realm of the longer waves. The radio-radar developments in the long wave lengths and the explosive energies of the gamma rays in the ultra-short wave lengths have created a new culture—the atomic civilization. In a few short decades the exploration of the entities Energy and Matter has changed the way of human life, has deeply affected man's social philosophy.

It is one thesis of this essay that these scientific discoveries, and the techniques built upon them, may have destined the older philosophies and creeds to substantial readjustment; they point to the possibility of radical modifications of some of the basic tenets. This idea is indirectly elaborated in later chapters.

The radiation spectrum is involved in many other outstanding developments, such as (1) the penetration of the ozone layer in the earth's atmosphere by the war-inspired rockets; (2) the modern alchemy of transforming one atom into another through bombardment with high-energy photons and high-speed electrons, protons, neutrons, and other corpuscles; and (3) the fission of heavy atoms and the fusion of lighter ones in the interest of providing atomic energy for beneficent peace and maleficent war.

Some of these items we must later sort out from the abundance provided by the electromagnetic spectrum and examine them in the interest of present interpretations in Cosmography and for future predictions. They should help in the placement problem.

Lesser Tables for Cosmography

The four major tabulations can advantageously be supplemented by a few other summarizing aids. Some of them are embodied in the subdivisions of the major tables—for example, the types of galaxies and the kinds of fundamental particles. Five relatively small, useful tabulations are the following:

(1) *The planets of the solar system,* their years, days, distances, sizes—all of which are significant in the consideration of the origin of the earth.

(2) *The major phyla of animals and plants;* all of them are the terrestrial descendants of sunshine and the primeval "thin soup" of the shallow seas.

(3) *Classes of mammalia,* from whales to bats to cows and the anthropoids.

(4) *The sequence of stellar spectra,* a color and temperature progression from bluish hot Rigel in Orion to yellowish Canopus and the sun, and down the temperature scale to reddish Betelgeuse and Antares.

(5) *Animal societies,* a series from the salmon with her eggs, to the monogamous robin, to the buffalo herd and human society, and on to the social integration achieved by honey bees and fungus-growing ants, which is the ultimate, perhaps, in social organization.

Summary of Orientations in Space, Time, Matter, and Energy

In concluding our review of the placement problem, let us see just where we stand with respect to the basic entities. We are dealing of course only with the material universe. I hardly see how we can locate ourselves in what might be called the "stream of thought," or find our place in some mystical spiritual category. We shall therefore summarize only in terms of our spot in time, space, energy, and matter.

(1) Time: Obviously in time we are precisely between the past and the future. Concerning the future we can extrapolate a little, but cautiously. As far as planets, stars, and galaxies are concerned, we see clearly no end to the material universe; we can only guess on the authority of incomplete theories. Concerning the past, the indications are clearer that somewhere between five and fifteen billion years ago there was an epoch T_0, of extraordinary significance in the history of our physical world. We believe there was a specifiable creative moment (or epoch) in the past, but no comparably specifiable moment in the future. We accept tentatively a finity in the past operation of an evolving universe, an infinity ahead.

Unless we deny T_0 and assume that there was no "creation in time," no real beginning of the dust- and star-populated Metagalaxy, no start of the expanding universe, we must conclude that we are relatively young in time. Our days are not near the end of the world, nor even midway. The hydrogen fuel that heats the stars is very abundant. We—the galaxies, stars, organisms—are just getting under way. Our 10^{10} past years are brief, negligible of course compared with a future of eternity.

There is, to be sure, an alternative hypothesis. It holds that the explosion of the primeval superatom that contained everything is illusory, and that the past is "just as infinite" as the future. On this rather tentative hypothesis, which tastes a little of theology and ancient dogma, there must be a continuous creation (emergence) of matter out of nothing to make up for that which, because of the expansion of the material universe, is lost "over the rim of the world." The primeval atom theory is the suggestion of Canon Lemaître; it is certainly consistent with many observations. The "continuous creation" hypothesis is associated chiefly with the names of Jordan, Bondi, Gold, and Hoyle; it has yet to get the observational backing to match its esthetic appeal. For the present we can safely accept much past duration, and as much or more in the future.

(2) Space: We are more easily located in the size category. It happens that man is just about as much larger than a hydrogen atom as the sun is larger than man. Geometrically, as we put it, we are in the middle register in the series of material bodies—that is, $\dfrac{\text{Star}}{\text{Man}} = \dfrac{\text{Man}}{\text{Atom}}$, and this is roughly true whether we are measuring in grams of mass or in centimeters of diameter. Man's location in space among the stars and galaxies is discussed in Chapter 7.

(3) Energy: To place ourselves somewhere in the energy table has practically no meaning. We could compute the amount of energy represented by the masses of our bodies and compare it with the energies represented by the masses of stars and atoms. But that comparison has already been done in effect in finding our place in the organization of material systems. We could locate ourselves in a vague way through indicating the energy that we command, which seems to be tremendous compared with the resources of our forefathers. Now we have fuel-fed dynamos; we have hydroelectric plants, and recently we conquered at least a part of atomic energy. But if we add all these terrestrial sources together and claim that they represent our standing in the energy category, they would be as nothing compared with a moment's radiation from an average star. It has been estimated that one substantial earthquake, which is neither man-made nor man-controlled, is energetically equal to a thousand atomic bombs; and, energetically speaking, a fair-sized solar prominence makes our "citybusters" dwindle to firecracker dimensions. In short, in the total cosmic energy operation and potentiality, man and his works are of minor consequence.

(4) Matter: By taking a vain and generous view we can claim a much better world position in the category of

matter. Negligible and incidental though we may be in space, time, and energy, we do have the distinction of sharing a wide variety of chemical atoms with the greats of the universe—with inanimate planets, stars, galaxies, and cosmic dust. Man is in a sense made of star-stuff. Important in his composition are a score of the elements found in the earth's crust. Some of the chemical elements are abundant in his body; others appear only as traces. The most prominent atoms in the make-up of animal bodies (mammalia) are the following, with an estimate of approximate percentages:

Oxygen	65%
Carbon	18
Hydrogen	10
Nitrogen	3
Calcium	2
Phosphorus	1
Others	1

In the atmospheres of the sun and sunlike stars a current theory suggests the following distribution of matter:

Hydrogen	81.76%
Helium	18.17
Oxygen	0.03
Magnesium	0.02
Nitrogen	0.01
Silicon	0.006
Sulphur	0.003
Carbon	0.003
Iron	0.001
Others	0.001

And by another interpretation of astrophysical evidence and theory the sun's composition is this:

Hydrogen	87.0%
Helium	12.9
Oxygen	0.025
Nitrogen	0.02
Carbon	0.01
Magnesium	0.003
Silicon	0.002
Iron	0.001
Sulphur	0.001
Others	0.038

In the earth's crust, including air and oceans:

Oxygen	49.2%
Silicon	25.7
Aluminum	7.5
Iron	4.7
Calcium	3.4
Sodium	2.6
Potassium	2.4
Magnesium	1.9
Hydrogen	0.9
All others	1.7

But of the earth as a whole, which includes the hypothetical iron-nickel core, we have the following estimate of composition:

Iron	67%
Oxygen	12
Silicon	7
Nickel	4
Others	10

All of the many human-body elements are of course in or on the crust of the earth, and most if not all of them have also been identified in the hot stellar atmospheres. No atomic species is found in animal bodies that is not well known in the inorganic environment. Obviously man is made of ordinary star-stuff and should be mighty proud of it.

In one respect animals and plants excel the stars. In the complexity of their molecules and molecular aggregates, living organisms transcend the atomic combinations of the in-animate world. The sun's hot atmosphere and also the solar interior are found to be relatively simple in chemical structure when compared with the organic chemistry of a cater-pillar. For that reason we are able to understand stars better than the larvae of insects. The former operate chiefly under the gravitational, gaseous, and radiation laws and are sub-ject to the consequential pressures, densities, and tempera-tures. The organisms are hopeless mixtures of gases, liquids, and solids—hopeless, that is, from the standpoint of our work-ing out for them neat and complete mathematical and physico-chemical formulae. The astrophysicist has a simple job com-pared with the demands on the biochemist.

★ 4 ★

An Inquiry Concerning Other Worlds

~~~~~~

Before we can propose ourselves and our destiny as significant concerns of the universe, we should turn our attention
to the possible existence and general spread of protoplasm
throughout stellar spaces and cosmic times. We can no longer
be content with the hypothesis that living organisms are of
this earth only. But before we ponder on the life spread, we
should inquire into the prevalence of suitable sites for biological operations. The initial question is not whether such sites
are presently inhabited. First we ask: Are there other habitable celestial bodies—bodies that would be hospitable if life
were there? No field of inquiry is more fascinating than a
search for humanity, or something like humanity, in the
mystery-filled happy lands beyond the barriers of interstellar space. But are there such happy lands?

Other Stars, Other Planets

It is generally admitted by practical people that we exist.
Extremely few cogitators on this subject pretend to a sus-
picion that we do not, that it is all a dream, an illusion, a
complicated fancy. Let us go along with the majority and

accept your existence and mine, and that of the physical world around us. And to simplify matters as a preliminary to discussion, let us say that the nonphysical world, if any, also exists. Around those words "nonphysical," "if any," and "exists" many battles could be fought, but the weapons would be mostly words, not ideas.

Since we exist on an earth where more than a million other kinds of animals are enjoying (or suffering) the same experience in biochemical evolution, we naturally meditate on the nature of this operation called living. We see a great variety of life forms and extreme diversity in living conditions and note also the wide adaptability of man. Naturally we ask: Are the likes of us elsewhere? The question is directed sometimes to the pastor or to the philosopher, but usually to the astronomer, and on behalf of astronomy I shall venture a reply; but in this chapter we consider chiefly the antecedent question: Are planets like ours elsewhere?

Human bodies are constructs of commonly known chemical elements, and nothing else. We have tabulated in the preceding chapter the principal atoms of the bodies of animals, and we must remember that in chemical composition humans are decidedly animal. The element oxygen accounts for about sixty-five per cent of our bodies; eighteen per cent is carbon, ten per cent hydrogen, three per cent nitrogen, two per cent calcium, and another two per cent includes silicon, phosphorus, sodium, sulphur, iron, and a dozen other elements—all common to the crust of the earth and to the flames of the sun. The percentages vary somewhat from rat to leech, from watery octopus to crusty coral. There is more than average calcium in the bony vertebrates, more silicon in the brachiopods, more H_2O in the jellyfish; but all animals employ all the common atoms. The elements uncommon to the rocks, like gold, platinum, and radium, are also uncommon to man.

No Life on the Stars

The stars are composed of the same stuff as that which constitutes the sun and is found in the earth's crust. They are built of the same materials as those that compose terrestrial organisms. As far as we can tell, the same physical laws prevail everywhere. The same rules apply at the center of the Milky Way, in the remote galaxies, and among the stars of the solar neighborhood.

In view of a common cosmic physics and chemistry, should we not also expect to find animals and plants everywhere? It seems completely reasonable; and soon we shall say that it seems inevitable. But to demonstrate the actual presence of organic life in other planetary systems is now impossible for us because the stars are so remote and we, as earthbound searchers of the sky, are yet too feeble in the face of stellar realities. To establish, however, through statistical analysis the high probability of planets suitable for living organisms is not difficult. A statistical argument, as a matter of fact, is more convincing than would be a marginal observation.

It will clarify the discussion if we start with two routine reminders: (1) by life we mean what here we terrestrials recognize as life—a biochemical operation involving carbon and nitrogen and making use of water in the liquid state (Other kinds are imaginable; e.g., one where silicon replaces carbon, or where sulphur's participation is like that of oxygen. Such is imaginable, but unlikely); and (2) Mars and Venus are therefore the only other planets of our solar system that are at all suitable for living organisms. The evidences are good that Martian life is low and lichen-like, if it exists at all, and the surface of Venus is an unsolved problem, with the odds against living organisms because of the lethal chemistry of the atmosphere.

Among the many definitions of life is the cold rigid version: "material organizations perpetuating their organization." The emphasis is on "perpetuating." We might better put it: "the perpetuation by a material organization of its organization." The definition can quite properly refer both to individuals and to species, and also to societies. They are all alive. They all die, if we suitably define death. The lively deathless atoms of our breath and bodies, however, are not, in this defining, alive.

Life is tough, tenacious, and also persistent when we give it time to adjust to varying environments. We find it in geysers and hot springs. Some flowers bloom under the snow. Both plants and animals on occasion endure for long periods on hot deserts. Some seeds can withstand desiccation and extreme cold indefinitely. Life as we know it on the earth has wide adaptability; but there are limits, and one of these limits is the heat and radiation near a star's surface, where the molecules constituting protoplasm would be dissociated.

In our consideration of the spread of life throughout the universe, we must therefore immediately drop all thoughts of living organisms on the trillions of radiant stars. The flames of the sun are rich in the lively atoms of oxygen, carbon, hydrogen, nitrogen, and calcium—the principal constituents of living matter—but physical liveliness and organic livingness are quite different behaviors. At the surfaces of some of the cool stars, like Antares and Betelgeuse, and in the cooling sunspots, we find a few familiar molecules, in addition to the scores of kinds of atoms; but there is nothing that is as complicated and tender as the proteins—those molecular aggregates that underlie the simplest life. And of course the stars harbor no water in the liquid state.

No Life on Meteors or Comets

The stars are out of it, therefore, and they probably represent more than half of all the material in the universe. Most of the rest is believed to be in the form of interstellar gas, with a bit of dust. The dust is of the sort that shows up as meteors, when in collision with the earth's atmosphere, and appears also as remote dark nebulosities that interrupt and make patchy the glow of the Milky Way. No life exists on these minute meteoric specks, or on the relatively larger meteorites, and for several reasons. Among the reasons: (1) the masses are too small to hold gravitationally an atmosphere (even our moon cannot maintain the oxygen and carbon dioxide necessary for breathing animals and plants); (2) moreover, the meteors out between the stars are too cold for liquid water; and (3) they are all too unprotected against the lethal ultraviolet radiation from hot stars.

How about life on the comets? The same general argument holds as for meteors and meteorites, since the comets are simply assemblages of dusty and fragmented meteoric materials, infused with escaping gases. In addition, most of the large comets of the solar system are, when brightest, too near the sun for living organisms and the rest of the time too frigidly remote in the outer parts of their orbits.

Only Planets Are Habitable

In our search for life we are therefore left with the planets, and those on which it can occur and survive must be neither too near their stars nor so remote from them that the cold is unrelieved. They should not be too small to hold an oxygen atmosphere, unless we are content to settle for primitive anaerobic life. (A few types of low organisms thrive in the absence of elemental oxygen.)

The life-bearing planets must also have nonpoisonous atmospheres, salutary waters, and agreeable rocky crusts; but given time enough, organisms could no doubt become adjusted to environments that would be poisonous and impossible for life such as that now developed on the earth.

Finally, the propitious planets that are suitable in size, temperature, and chemistry must also have orbits of low eccentricity. Highly eccentric orbits, like those of most comets, would bring their planets too near the star at periastron a part of the year, and then too far out at apastron. The resulting temperature oscillations would be too much for comfort, perhaps even too much for the origins and persistence of early life. Also, in the interests of avoiding too great differences in temperature from night to day, it would be best if the planets rotate rapidly and their rotational axes be highly inclined (as is ours) to their orbital planes.

With the foregoing requirements in mind we ask if there are many really suitable planetary systems, and the companion question: How are planets born?

Genesis—A Twentieth-Century Version

In the beginning, as they say, was chaos. Or at any rate, soon after the explosive beginning of the expanding universe there was chaos, if we accept the theory of the Primeval Atom as proposed by Canon Lemaître—a theory that visualizes the original assemblage of the matter and energy of the whole universe in one body, a single superatom. In those chaotic early times, some five to ten billion of our years ago, the average material density was of course very high; the stars were near together; many galaxies interpenetrated, if galaxies at that time existed as organizations, and if not, the forming protogalaxies overlapped. Collisions and secondary explosions must have been frequent in those crowded, chaotic times. Masses of flying gas in the cold of space quickly liquefied, solidified, cooled into planet-like bodies, with a wide range of sizes. Shattering and exploding bodies produced dust grains and gases from which later stars were born. All that action seems logical, if we accept the hypothesis of a tight little universe before the cosmic expansion had spread it out. Moreover, better hypotheses are hard to come by.

Let us proceed to sketch in more detail this particular hypothesis relative to the earth's birth. The larger products of the explosive expansion of the superhot Primeval Atom would remain gaseous. If their masses were not too large or too small they became eventually luminous stars. Those that were too extended to hold together as single stellar bodies in equilibrium between the pulling in of gravitational force and the outpushing of radiation pressure and gas diffusion, plus the centrifugal action of bodies when in rapid rotation— those oversize bodies became double or multiple stars, or clusters of varied population.

The Lilliputians

The masses that were much smaller than our sun but larger than our planets—and there must have been billions of them— provide a mystery, an unexplored element of the stellar universe. Our astronomical records show none of them, but our time is brief on the cosmic clock. They would have mass enough to contract into stable permanent bodies with dense atmospheres, but not mass enough to shine so effectively that they could be seen unless very near. The whole range of sizes, from those of 1/50 the solar mass, which we might call dark Lilliputian stars, to those 1/500 the mass of the sun, which we might call Brobdingnagian planets, could be represented by countless sidereal bodies. They may be more numerous than the recognized stars.

The largest of the Lilliputians would shine faintly in infrared light, and could be detected if near at hand; but mostly they are lightless wanderers. Eventually scientific techniques may enable us to detect them. For example, by radio signals if such unseen bodies have violently electrical atmospheres or register themselves through excessive volcanic activity. One of them may some day drift into our planetary system, being first manifested by its reflection of sunlight or through perturbations of the motions of our outermost planets and comets. It would be very instructive to meet up with one of these intermediate products of an early explosive epoch. Their existence can scarcely be denied. Their abundance is uncertain, but probably high. Theories of planetary origin other than the one we are now sketching would also make probable the Lilliputians.

Primeval Chaos and the Clean-up

Whether or not there was in the beginning a single primeval "atom" that contained all, there can be little doubt that the now dispersing galaxies with their billions of stars were densely crowded together in the remote past. It is highly significant that the age of the earth's crust, measured by the radioactivity in the rocks, is much the same as the measured age of the expansion.

To go on with the sketch, let us suppose that one particular emerging star (our protosun) was densely surrounded with a miscellany of the debris of the original or of some subsequent explosion. These cooling secondary bodies would be at least loosely in the gravitational control of the star. Among the sun-circling bodies (protoplanets) we further assume that there were one or two so large that they participated effectively from the beginning in the control of the lesser bodies. If two of these large secondary bodies, which we may name proto-Jupiter and proto-Saturn, were moving in approximately the same plane and same direction around the protosun, they would be in a position to govern the motions and dictate the future of the lesser bodies. Some of the small planetary bodies —perhaps most of them—would be moving in orbits too elliptical for safety; they would be either engulfed at perihelion by the sun or by the two giant planets when in the outer parts of their orbits, or perturbed entirely out of the solar system into interstellar space.

In the billions of years that this tripartite mill of the gods has been working, the retrograde bodies, i.e., those circling the sun in the direction opposite to that of the large planets, would be disposed of through gravitational perturbations. In fact the flimsy comets are handled in this fashion in these late days of the mopping-up. The directions of their motion are sometimes substantially altered and the eccentricities of their orbits greatly changed. A few are "captured" into smaller orbits. In the long past many must have been ejected forever from the solar system by the larger planets. For when distance diminishes the sun's gravitational power, a massive planet can control the destiny of any comet whose orbit brings it near, and one of the possible edicts is ejection.

According to the proposed hypothesis of Original Chaos and Slow Mopping-up, the protoplanet earth was one of the favored fragments of the catastrophic operation that gave birth to the sun and its accompanying debris. Its orbit was safely circular, or nearly so; its spacing from neighboring planets (Venus, Mars) was such that perturbations were not serious.

At the dramatic time when the solar system was forming, other stars, some of them near the sun, were undergoing a similar experience. It would have been an exciting time to be a beholder of the young sky, rich as it then was in comets and meteors and with unseated planets on the loose.

Even if the earth's original orbit and those of the other surviving planets were considerably elongated in the early days, the primitive interplanetary medium of dusty debris, through which the planets circled about the sun, would round off the orbits. This medium of dust would resist the planetary motion; it would tend to lessen the eccentricity of the orbit and thus insure its safety from collision and engulfment. Most of the orbital inclinations (tilts) in the present solar system are similar to those of Jupiter and Saturn—a result, the hypothesis would suggest, of the material impacts and gravitational pulls in the early and long continued times of adjustment from chaos to order.*

The foregoing suggestions as to the origin of the sun's planets, and of other planets around other stars, were proposed long ago by the writer in a time of desperation. The two prevalent theories of planetary origin—nebular contraction and tidal disruption—were at that time in trouble with the observed facts, and with the accepted notions about the character and behavior of interstellar dust. Facts have been the No. 1 enemy of cosmogonic theories. If we did not know so much, we would have less to explain. The above chaotic-origin theory gets around many of the serious difficulties confronted by earlier hypotheses that have tried to explain all the major regularities in the solar system; but it also has troubles, or, rather, insufficiencies. It needs bolstering here and there with protective subhypotheses, as do all the others.

---

* A better statement would be: "times of adjustment from *apparent* chaos to order," for in this physical world there is no *real* chaos; all is in fact orderly; all is ordered by the physical principles. Chaos is but unperceived order; it is a word indicating the limitations of the human mind and the paucity of observational facts. The words "chaos," "accidental," "chance," "unpredictable," are conveniences behind which we hide our ignorance.

The major assumptions back of the chaos theory are the two following:

(1) A catastrophic origin of stars (including our sun) some few billions of years ago is assumed to have filled space with gaseous, liquid, and solid debris of all sizes, and gravitational control over a limited section of space was the natural power of the most massive fragments; in our case the control was vested in the sun.

(2) In that primeval aggregation of dust, gas, and planetoidal fragments, dominated by the protosun, one or two large protoplanets are postulated to have circled the all-controlling sun in approximately what is now the mean orbital plane of the solar system.

The rest of the procedure follows naturally—the sweeping-up (or out) of most of the debris and the ordering of the minor pieces into regions and motions of safety.

We omit further elaboration of this hypothesis and simply suggest that its insufficiencies include the poor accounting for the remarkable spacing of the surviving planets (this similarly affects other hypotheses) and the almost total absence of high inclinations for the orbits of bodies in the solar system, although long-period comets pretty well ignore the common orbital plane and some of the asteroids, between Mars and Jupiter, still have high inclinations.

A serious thrust at this hypothesis would be made if we could prove that stars arise not full size (or nearly full size) from the primeval exploding atom or a later (supernova?) outburst but emerge through the slow condensation of interstellar nebulosity. Also, a protecting subhypothesis, or abandonment, would be necessary if we could assure ourselves that the earth actually was built up gradually over a very long time by the slow accretion of interplanetary matter. Nevertheless, the hypothesis has the advantage that out of explosions, and what we call chaos, almost any original arrangement of materials and motions can be assumed. The present puzzling distribution of orbital and rotational speeds (the angular momentum problem) would be irrelevant. As a working hypothesis to show one way in which planetary systems can be formed, the chaos theory is worth preserving, at least until some comprehensive alternative is developed and widely accepted as the only reasonable Genesis.

The foregoing suggestion has been presented in some detail because if such a theory eventually prevails it would carry with it, as an important corollary, the implication that planetary systems may be just about as common as the stars and that conditions suitable for life permeate the cosmos. And then the questions "Are we alone?" and "Are we a unique biological construct of the universe?" could be answered negatively and emphatically. But there are many other theories of the earth's origin and in the interest of our concern about the distribution of biological habitats we should name those that are more or less convincing.

Many Methods of Planetary Origin

The supernatural deities of various sorts, rather than "accident" or astrophysical operations, were in ancient times given credit for the origin of the earth; but also the assumption that it always had existed was not uncommon. Many of the rationalizations, developed to account for the origin of the inclusive system of sun, planets, satellites, asteroids, comets, and interplanetary dust, are now wholly discredited. Some theories included the origin of the sun; others assumed its prior existence before the planets appeared. Most of the theories are of recent date—a natural consequence of the great accumulation of relevant scientific data in the past few decades, and of the increasing population of ingenious speculators. The new knowledge built up by the world's astronomers in the past forty years is many times that of all times before.

The following list of fifteen hypotheses, arranged in approximate chronological order, represents the thoughts of speculative scientists from Israel, Germany, France, Australia, India, America, England, Russia, Sweden, and Holland. This geographical distribution shows the wide spread of curiosity about man's physical place in the universe.

(1) The Mosaic cosmogony, and similar early religious doctrines.

(2) Nebular hypothesis, the famous long-enduring Kantian-Laplacian theory.

(3) Partial disruption of the sun by a comet, with the production of planets.

(4) Solar eruptions providing planet-building "planetesimals."

(5) Capture of the planets by the sun from space or from other stars.

(6) Tidal disruption of the sun by a passing star, providing gaseous filaments that condense into planets (variant of 4).

(7) Glancing collision of stars (variant of 6).

(8) Break-up of one component of a binary star by a third passing star.

(9) Explosive fission of the hypothetical protosun.

(10) Disruption of an unstable pulsing variable star (cepheid).

(11) Revival of the nebular hypothesis, bolstered by modern theories of dust and gas accretions.

(12) Electromagnetically produced condensations in a contracting nebula (variant of 2).

(13) Nova explosion in a binary system providing circulating planetoidal fragments.

(14) Revival of the hypothesis of cold planetesimals operating in a nebulous medium (combination of variants of 4, 11, and 12).

(15) Primeval explosive chaos and the Survival of the Conforming—my "hypothesis of desperation" outlined above.

All of these theories could be described in detail. Some of them overlap. A few have been rejected on the basis of obvious failings or because they are not complete hypotheses. For example, Nos. 1, 3, 4, 5, 9, and 10 are out. Nos. 2 and 6 are weak. This leaves Nos. 7, 8, 11, 12, 13, 14, and 15. We should observe that several different methods may be responsible for planetary origins. We do not need to search out only one method to the exclusion of all others.

The general conclusion at this moment must be that not one of the theories is entirely satisfactory. The best of them need further development. Many do not easily account for the following observed regularities and arrangements:

(a) The nine major planets revolve in the same direction around the sun.

(b) The sun and, so far as known, most of the planets and satellites rotate on their axes in this same direction.

(c) The inclinations of the planetary orbital planes are such that the over-all system (excluding comets) is exceedingly flat.

(d) The smaller planets, except Pluto which may be an escaped satellite of Neptune, are relatively near the sun; the greater planets are all from five to thirty times the earth's orbital radius from the sun.

(e) The satellite systems of Jupiter and Saturn have characteristics simulating those of the planetary system of the sun.

(f) Apparently the chemical content of the earth, probably also of the other planets, is similar to that of the sun, when allowance is made for escaping atmospheres.

(g) The distribution of angular momentum in the planetary system appears fatal to many of the hypotheses; the sun rotates too slowly, or the planets too fast to allow for a common origin—unless protective subhypotheses hurry to their assistance.

The foregoing theories, with two or three exceptions, can be classed as either catastrophic or as calmly accretional. In other words, the planets were born of violence or of slow building up through the accumulation of material. The former generally visualizes the earth, or at least the protoearth, as once hot from surface to center; the latter visualizes the earth's surface as never wholly molten, although, as the mass grew from accretion, the center naturally heated up and affected the outer strata.

What do these theories imply with respect to the prevalence of earthlike planets? If the contracting-nebula type of origin of stars is accepted, and the stars are held to result mainly from the condensation of cold clouds of gas and dust with the planets coming along as a by-product, then we must assume that planets like those we know, similar in mass, temperature, and chemistry, are the natural and common product of an evolving universe. The same high frequency would prevail for the Original Chaos and Sweep-up hypothesis.

We must always remember that our sun is a very ordinary sort of star. One hundred thousand of the brightest million stars are essentially identical with the sun. We detect this similarity by way of spectrum analysis (described briefly in Chapter 6), which tells us of the luminosities, masses, sizes, motions, and chemical compositions of the stars. We get the proportion, ten per cent, from large samples. Moreover, stars do not need to be like our sun to have habitable planets. If the star is hotter, the liquid water zone is farther out; if cooler, the livable planets must be nearer the stellar power plant.

The sun and the aforementioned 100,000 stars like it have

no exceptional position in the Milky Way system; they are in the outer part of what appears to be a typical large spiral galaxy in a Metagalaxy where there are thousands, perhaps millions, of galaxies of the same spiral type. The sunlike stars (most of them) also have a history dating back to those turbulent planet-breeding early times. The evidence is increasing, therefore, for an abundance of habitable planets.

There is, however, in all theories of origin one important deterrent to the universal formation and retention of planets with the suitable requirements for life. That hindrance lies in the common existence of numerous double and multiple stars. A century ago double stars were considered something of a rarity. With the increase of optical power and of skill in discovering binaries and multiples, the picture has changed. We now believe that forty per cent or more of the stars are in pairs or triples. Of the fifty-five stars known to be within one hundred trillion miles of the earth, only thirty-one are single stars and companions may yet be found for some of them. In a two- or three-star system, planets within the zone of liquid water are highly improbable. The gravitational rules are against it. Orbits would be unstable. We must accept as hospitable only single stars, and perhaps very wide doubles where a stable planetary orbit around one component might be permitted by the other star. Probably we should also exclude the highly populated centers of globular star clusters if we want long endurance for a system of planets.

Infrequency of Collisions

By taking the contracting-nebula hypothesis (say, No. 14), we would choose the one that probably is most favorable to the formation and preservation of planets. But suppose we accept instead one of the collisional hypotheses; not the cometary-collision suggestion, for the comets we know are relatively too small in mass to be a potent factor in stellar catastrophe. The collision of two or more stars might now be very unproductive of planetary systems because of the infrequency of such collisions. At the present time our sun and its neighbors, and this holds for all single stars not in the center of clusters or the nuclei of galaxies, are so remote from each other that collisions are well-nigh impossible. Twenty-five trillion miles separate our sun from its nearest known stellar neighbor, Alpha Centauri. Any given star, although it be moving at the average relative speed of twenty miles a second, would course around the galaxy for millions of years without collision or near approach. If we insisted that planetary systems could arise only through the collision or very near approach of mature stars in a fully developed galaxy like ours, we would need to subscribe to the belief that our own "accident" may have been unique—one time only in the whole galaxy—and we the offspring of that remote improbable event! We should then answer: "We *are* alone; we are the special care of whatever omnipotence is concerned in caring for rare accidents."

But two observations quickly weaken or defeat that conclusion. The first is that we do not limit our thoughts only to this galaxy of 100,000,000,000 stellar bodies that are always attracting but safely avoiding each other; we must consider the increase in the chance of collision somewhere provided through the circumstance that there are billions of other galaxies within our telescopic reach, and possibly trillions beyond our direct knowing. All of these systems must be considered when we examine the probability of life as a cosmic phenomenon. If only one galaxy in a hundred has had a planet-producing collision among its stars, there would be millions of such collisions.

The other and more potent contribution to the defeat of our isolationism (on the hypothesis that only collisions can produce planetary systems suitable for life) is the relatively new and well-established evidence that the Metagalaxy is expanding. The actual observation is that galaxies are receding in all directions from each other and that the average density of matter in metagalactic space is steadily decreasing. We need only turn that observation backward and conclude that yesterday the galaxies were closer together than today, that a million years ago they were still nearer each other, and that a few billion years ago they were all tightly packed together, intermingled, overlapping. They were far back toward the stage dramatically described as Canon Lemaître's "Primeval Atom." In the early days collisions and disruptive near approaches must have been millions of times more frequent than would be possible now. In the whole wide world of galaxies there must have been by one method or another innumerable planetary systems established—precariously at first because of the interference of other stars, but ever more safely as the universe expanded.

Billions of Planetary Systems

On the basis of our sampling census of stars in our galaxy, and our sampling of galaxy population out to the limit attainable by present telescopes, we can readily compute that there are more than 10^{20} stars in the universe, each one competent of course through radiation to maintain the photochemical reactions that are the basis of plant and animal life. Perhaps only a few per cent of these are single stars with planetary potentialities. Perhaps only a few per cent of these few developed in such a way (nebular contraction) or had such a suitable experience in the past (collisional) that they would now possess persisting planets. Perhaps only a few per cent of these that succeed in having stable-orbited persisting planets would have one or more at the right distance from the central star; and of these rightly placed planets but one per cent have an orbit of suitable circularity to maintain sufficiently equable temperatures. We could go on with further restrictions to a few of the few of the few, because nonpoisonous airs and waters are also required, and that particular activity of carbon, oxygen, hydrogen, and nitrogen that we call "living" must get started. We could by such restrictions reduce the number of stars with livable and actually "inhabited" planets to nearly a nothing.

All these restrictions, however, get us practically nowhere in isolating ourselves as something unique and special, for there are too many stars. Three undeniable factors have entered our consideration—the ordinariness of our sun which has accomplished the creation of life on this planet; the evidence of the universality of the kind of chemistry and physics we know here; and the existence of more than 10^{20} opportunities for life, that is, the existence of more than one hundred thousand million billion stars.

Let us look once more at large numbers and work this argument over again. Suppose that because of doubling, clustering, secondary collisions, and the like, only one star in a thousand has a planetary system. Personally I would think that one in fifty would be a better estimate, and many of those who believe in the nebular contraction theory of stellar formation would say that at least one out of ten stars has planets. But to be conservative, we say that only one out of a thousand has a planetary system, and then assume that but one out of a thousand of those stars with systems of planets has one or more planets at the right distance from the star to provide the water and warmth that protoplasm requires. In our solar system we have two or three planets in such an interval of distance. Further, let us suppose that only one out of a thousand of those stars with planets suitably distant has one large enough to hold an atmosphere; in our system we have at least seven planets out of nine with atmospheres.

That will reduce our suitable planet to a one in a billion chance.

Let us make one other requirement of our suitable planet: the chemical composition of air and water must be of the sort that would develop the naturally arising complex inorganic molecules into the organic. Perhaps that happens but once in a thousand times?

Assembling all four of the one to a thousand chances (all grossly underestimated, I believe, but in the effort to establish our uniqueness in the world, and hence our "importance," we are making it as hard as possible to find other habitable planets), we come to the estimate that only one star out of 10^{12} meets all tests; that is, one star out of a million million. Where does that high improbability of proper planets leave us? Dividing the million million into the total number of stars, $10^{20} \div 10^{12}$, we get 10^8—that is, a hundred million planetary systems suitable for organic life. This number is a minimum, and personally I would recommend, for reasons given in Chapter 5, its multiplication by at least a thousand times, possibly by a million.

To state a conclusion: The researches of recent times have enriched and clarified our concepts of habitable planets. Through discovering the true stellar nature of the spiral "nebulae," through the sounding of star-and-galaxy populated space to such great depths that the number of knowable stars rises to billions of times the number formerly surmised, and through the discovery of the expansion of the universe with its concommitant deduction that a few billion years ago the stars and planetary materials were much more densely and turbulently crowded together than in the present days of relative calm, we have strengthened our beliefs with respect to the existence of other "worlds." The present concept includes the identifying of our own world as the surface of planet No. 3, in the family of a run-of-the-mill yellowish star, situated in the outer part of a typical galaxy that contains billions of typical stars—this "home galaxy" being one item in an over-all system, the Metagalaxy, that numbers its galaxies in the multibillions.

So much for the peripheral position of our planetary system. So much for the commonness of habitable planets. In the next chapter we turn to the question of what habitable planets are in fact inhabited.

★ 5 ★

On the Hazards of Primitive Life

It is really of little importance to our main argument whether there are very abundant locations suitable for biology or only a million planets where sentient organisms live and react to their respective environments. Even this small number firmly establishes the need, in any approach to the interpretation of Nature, of including extraterrestrial life in our cosmic picture. Terrestrialism by itself is limited, useless, and now essentially dead. This viewpoint has come about, and is already gaining wide acceptance, through the impact of three important and relatively recent scientific discoveries. They are the great number of galaxies and stars available for energizing life, the crowded state of the universe we know a few billion years ago (which would result in the birth of numberless planetary systems), and the partial bridging, in the chemical laboratories, of the gap between complex but lifeless natural molecules and the simplest manifestation of continuing life. The discussion of man's Fourth Adjustment in Chapter 7 will spell out these three contributions.

Notwithstanding the little need of further emphasizing the wide spread of protoplasmic existence, it may be well to discuss in more detail the prevalence of life, the hazards it must face, and the possibility of other kinds than ours. The fact that life exists on this undistinguished planet, and arose here naturally, as we shall indicate in Chapter 9, is by itself nearly enough assurance that life is a cosmos-wide occurrence.

Early Protoplasmic Perils

The argument in preceding chapters maintains that at least one star in a billion will have with it a crusted nongaseous planet entirely suitable for living organisms. These starlit planets should be as satisfactory for life as is our planet which, with the help of sunlight, has in the course of time developed on its land and in its seas millions of kinds of plants and animals. We feel assured, from our extensive sampling of the population in distant regions, that more than 10^{20} stars exist; therefore these suitable one-in-a-billion planets are not at all scarce in the over-all stellar universe.

But it does not follow that just because a planet is suitable for life in having the proper air, water, warmth, and varied chemicals in solution it will of necessity harbor highly evolved living organisms. It does not follow that all these hundred thousand million life-bearing planets will have developed highly sentient beings, corresponding to or excelling the best our earth has done. For protoplasm has its perils, especially inexperienced protoplasm.

All living conditions might appear to be propitious on a planet and still nothing happen of high biological significance. That would seem to be extremely unlikely, of course, to one who believes that life is both inevitable and persistent when conditions are right, and the range of rightness is wide. But suppose that in some critical early epoch the air suddenly had so much of the free oxygen that it would burn out the tender molecular combinations striving toward biochemical viability. Or suppose that the cosmic rays from space were too strong or too weak, or too much blocked by atmospheric molecules for them to play a proper part in the initial energizing of the activities that were required for the combining of the methane, ammonia, water, and hydrogen of the primitive atmosphere* into the simplest amino acids. And conditions might have been adverse to the continuity of life once started. Doubtless on many planets for various reasons the most primitive organisms were stillborn.

On this planet life started in the early days of the expanding universe, and here it has succeeded in holding on and growing. Soon the vegetation began to take part in the evolution of the earth's atmosphere, helping to replace the dominant water vapor and the hydrogen of the primitive gaseous medium with the present abundant oxygen. The output of oxygen by plants is now partly balanced by the intake of oxygen by animals; the carbon dioxide exhaled by the animals (assisted in CO_2 production by vegetable decay, fires, and volcanic exhalations) provides the carbon essential for the carbohydrates of the plants.

Probably life got started also on Mars, Planet No. 4. The chemical make-up of the Martian atmosphere and the climatic conditions are considerably different from ours; but there the urge to evolve is as likely, or at least was as likely, as here on No. 3. The oxygen shortage, however, keeps Martian organisms (if any) in low estate. (In Chapter 9 the difficulties of getting born are further examined.)

If among the planets suitable for the higher biological developments there is only one in a thousand that has actually carried its protoplasmic operations to the level we enjoy, that numerical modification would still leave us, as shown in the preceding chapter, with at least 10^8 high-life planets—that is, with more than a hundred million domiciles of highly developed organisms. We must keep in mind that our sun is average, our stellar associations are not unusual, and our location in space is in no way so special that this particular planet

---

* See Chapter 9 for a report on other atmospheric compositions.

should receive preferred treatment and uncommon opportunities in a biochemical way.

We have cut down from more than 10^{20} to 10^8 the estimated number of stars that feed higher organisms. Many of the other stars may be free of long-continuing biological operations on their planets, or may support only low forms of life. The disqualifications we have suggested on the grounds of star-doubling, poor locations, bad chemicals, and so forth, have been ruthless. We have sought all ways to write off the existence of competition at the high life level. The search has been a gesture to those who harbor the wish or thought that Homo may have the distinction that goes with complete uniqueness.

Personally, the writer would multiply the favorable chances for life by a million, accepting the existence of at least 10^{14} high-life planets. There are two principal reasons for this increase by a million times. The first is that we have probably grossly underestimated the number of stars in our universe. The second is that other kinds of life may exist, not only that based on carbon compounds.

The theoretical considerations by Eddington and later by others suggest that there are not less than 10^{79} fundamental particles (electrons, protons, neutrons) in the universe. The total mass is therefore more than 10^{55} grams,* that is, $10^{79} \div 10^{24}$. Taking as a standard star a mass one-half that of the sun, thus allowing for the vast number of dwarf stars that current researches reveal in the solar neighborhood and which probably exist elsewhere, we have $10^{55} \div 10^{33} = 10^{22}$ stars. This consideration increases the number of life chances by a hundred times over the conservative number given above.

That the number of stars mentioned earlier, 10^{20}, is much

---

* There are approximately 1.2×10^{24} fundamental particles (protons and electrons) in a gram of matter (counting a neutron as two particles).

The mass of the sun is 2×10^{33} grams. Each "standard" star therefore contains about 10^{57} particles.

too small an accounting can be inferred also from observation. Our sampling of space shows that at least a billion galaxies are within four billion light years. If they are on the average only one-tenth as rich in stars as our own galaxy, there must be $10^9 \times 10^{10} = 10^{19}$ stars now within our present sampling. A reach to only ten times our present probe would run the number of stars to something like 10^{22}. And that extension of reach is not asking too much of the future. Between 1915 and 1930 we increased the length of our celestial surveying rod by nearly a million times and therefore increased the explorable volume by the cube of that number.

Other Kinds of Life?

The second argument for a substantial increase in our estimate of the number of planets satisfactory for high organic forms is of a quite different character. It concerns biochemistry rather than statistical astronomy. Omitting details, it can be put as follows.

The life we know and which we have loosely defined on earlier pages is essentially that which, in our mixture of precocious grasp and profound ignorance, we would recognize, and designate as life, on the surface of any planet we should visit. By surface, we include the atmosphere, oceans and other surface waters, and the land. Also included are the depths of the oceans and the depths of soils that cover the rocks. In all these terrestrial locations, life is chemically much the same. Carbon compounds prevail; so we usually say that our life is based on carbon, the sixth element in the periodic table.

In its commonest form, a carbon atom is made of six protons, with their six positive unit charges, six neutrons that add weight but add no charge to the nucleus, and six nucleus-surrounding electrons, with their six negative charges balancing the positive total charge on the nucleus. The six electrons of the carbon atom, according to a convenient atomic model, are in two shells—the two inner electrons in the so-called K-shell and the other four in the outer L-shell. The electron structure is written 2-4. The next element, nitrogen, has a 2-5 arrangement, and oxygen 2-6. Because of their electron shell structure, these elements, particularly carbon and oxygen, are easily combined with each other and with hydrogen and the other elements found in organisms.

In the same vertical group of the periodic table (page 36) is another very common element; in the earth it is nearly two hundred times as common as carbon. It is silicon, which constitutes nearly a quarter of the earth's crust, and in its usual molecular combination with oxygen (sand) makes up about three-fourths of all rock material. Its electron structure is 2-8-4. The outer four electrons, in the M-shell, make of this atom another ready joiner with hydrogen, nitrogen, and oxygen. Like carbon, it can appear in compounds that are gaseous, liquid, and solid. But CO_2 is gaseous at room temperatures; SiO_2 is gaseous only above 2500° C.

These two elements show some other differences, and many similarities, in their combinations with other elements. Here we need only remark that life based on silicon compounds rather than carbon compounds would be unlikely, but it is a possibility that we must not overlook. The type of life might be one that we would not readily recognize, since it is not at all probable that the same sort of complex organic molecular aggregations would naturally arise, or that operations paralleling the photosynthesis we know would or could occur, or that the metabolism of a silicon-based organism would be recognizably like that of our carbon-oxygen-hydrogen organisms.

Other elements (like sulphur, electron structure 2-8-6) have been suggested as possible substitutes for oxygen in biochemical development. Certainly we must not limit the possibilities of biological life only to planets with atmospheres, waters, and soils like ours. On the presumption that other viable chemistries may prevail, the number of planets hostile to organic life might thus be considerably reduced.

Adjusting to Unearthly Environments

The gradual adjustment to chemical and climatic conditions that are far different from the kind that terrestrial organisms now experience, must be allowed for. A situation that would today be lethal for us might have become tolerable through slow adjustment. Poisons taken in small but increasing dosage sometimes lose their toxicity. If accustomed to ultraviolet radiation over long ages, we could probably stand much more of it than we now endure. Indeed, it is likely that early life on the earth was subjected to very strong ultraviolet radiation, since the present ozone (O_3) barrier some twenty miles above the earth's surface presumably is largely a by-product of the gradual evolution of the atmosphere as vegetation began to release oxygen on a large scale.

Biological adjustment to various physical, chemical, and climatic conditions could widen considerably the life zones around a star. Homo sapiens, for example, is a most adaptable animal and now successfully swarms over the whole earth with the protection of clothing, skin pigments, parasols, and domicile heating. He can go to high altitudes of low pressure with oxygen tanks; and go into the heavy atmospheric pressure of mines and the water pressure of ocean depths when outfitted with the gadgetry that helps to simulate the conditions of his natural habitat. Many animals and plants can do better than man, thanks to long and gradual adjustment. Insect life in hot springs, lichens in the polar zones, marine forms at enormous pressures in the deep seas all testify to the breadth of adjustment and encourage the belief that biological evolution, even if limited to the carbon-water-nitrogen chemistry, can be expected on planets differing considerably from the norm of our present temperate and torrid zones.

Perhaps One Star in Every Million

The foregoing considerations of the climatic and physical extremes that nevertheless support life, and of the possibility of organic origins and evolutions based on other chemical operations, have led to the second revision of the numerical probabilities of life throughout the universe, and justify the surmise that an enhancement by a factor of a million is reasonable—namely, that we should contemplate at least 10^{14} planetary situations for life at our level of sentiency. In other words, we surmise that at least one star out of every million supports some kind of high-level protoplasmic operation on one or more of its planets. Many, but not necessarily all, of these 10^{14} planets probably have the plant-animal interdependence in which we ourselves participate.

The carbon and oxygen exchange is in a sense the breath of life—the indispensable inhaling of carbon dioxide (by plants) and oxygen (by animals), and the exhaling of oxygen (by plants) and carbon dioxide (by animals). It is a cosmic symbiosis. In the total absence of animals on a planet, vegetation that depends on photosynthesis might soon find the carbon dioxide in short supply; the plants would need to depend for their carbon mostly on erratic volcanos, natural fires, and the products of their own decay. In the total absence of plants, animals would of course promptly starve for lack of energy; they would in fact never have evolved. As it is, we live together, nutritionally, in happy symbiosis. We animals use the plants for fixing carbon and releasing oxygen, that is, for supplying food and breath; the plants use us as one of the sources of their carbon dioxide, and as fertilizer. It is a natural barter economy.

Life on Planet X

Exactly where these other life-bearing planets are we cannot now say; perhaps we never can, lost as they are in the glare of their stars, isolated as we are in space, and equipped with sounding apparatus that is still, we hope, primitive. Although not seen or photographed, those planets are deduced as statistical probabilities. There must be at least 100,000 of them in our galaxy, if we accept the frequency the writer prefers; but only one in every dozen or so galaxies if we accept the most ruthless shaving of the probabilities.

Nor can we say what kind of organisms inhabit these other worlds. Are they only plants, animals, and the simpler of noncellular organisms (protista)? Or are there perhaps other highly developed animate kingdoms, neither plants nor animals, nor in between? If a mature tree would at times pull up its roots and stroll away to a more nutritious locality, or if an animal should give up its mobility at times, put down roots and feed itself through the soil and the material provided by photosynthesis, we would be astonished. But stranger operations are already known to occur in the lower life of the earth's biology. Chlorophyll, carotin, and xanthophyll may not be the only means of capturing the sun's radiant energy. Our yellow star radiates the wave lengths that we have grown to need. Redder and bluer suns would generate and cooperate with life forms that prosper in radiation emanating from redder and bluer sections of the spectrum. Their chemical energy converters may not even resemble our chlorophyll.

Although the life on what we shall call Planet X, an unidentified high-life planet, is a matter for loose conjecture, we should naturally expect it to resemble in many ways some of the myriad life forms on the earth. It is amazing that there are biochemical properties and fundamental growth behaviors

that are the same for thousands of kinds of plants and animals
that superficially and in bulk differ greatly. Clovers and
sequoias have common characteristics in the flow of sap, in
the structure of their stems, and the function of their roots
and leaves. In mice, whales, and men, the heart, lungs, and
brains are similar in cellular structure and in basic operation.
Parallelism, both in evolutionary steps and in attained ends, is
frequent among the earth's animals, suggesting that such steps
are inevitable in biochemical evolution. The properties of
atoms and molecules may be such that growing molecular systems and evolving organisms are restrained to develop along
specific lines in accord with specific patterns. We might call
this an inborn orthogenesis growing out of the properties of
organic molecules. We should in that case expect much the
same biological operations on Planet X as here, with comparable end results.

The terrestrial social insects, for instance, provide a notable
example of parallel development. In many important characteristics the agricultural ants and some species of termites are
alike. The two insects are similar in body size, caste system,
and the existence among them of sterile workers and egg-lay-
ing queens; also in practice of fungus culture, casting of the
queens' wings, detailed care of young, and in parasite tolera-
tion. Despite these characteristics of astonishing similarity,
the termites and ants are not at all closely related. They are
phylogenetically as remote from each other as man is from the
whale or the bat. But basic chemistries and fundamental
physical laws apparently decree that the elaborate organizations of termites, social bees, social wasps, and ants, in their
thousands of species, must travel much the same road toward
their integrated societies.

A mixture of pure chemical elements will always under the
same physical conditions produce the same result, whether it
be an odor, an explosion, a color. Perhaps we should expect
that a mixture of starshine, water, carbon, nitrogen, and other

atoms, when physical conditions are fairly similar, will everywhere produce animals that are much alike in structure and operation and plants that have certain standard behaviors, notwithstanding great morphological differences. If we should visit a planet essentially identical with ours in mass, temperature, age, and structure, we would probably not find the biology queer beyond comprehension. We might find it no more peculiar than we would find the biology if we were transported on our own planet into Carboniferous times, or taken back just 150,000,000 years when the great lizards ruled the land and sea, and the birds, mammals, and flowering plants were not yet far developed.

Therefore we surmise that the biology on Planet X and Planets Y, Z, and so forth, might have much in common with the living forms on Planet Earth just because the carbon compounds will have it so, and because the same chemistry and the same natural laws prevail throughout the universe we explore.

⋆ 6 ⋆

Rainbows and Cosmic Chemistry

Notwithstanding many important contributions from biochemistry, including those to which we refer in Chapter 9, the mystery of life is not wholly resolved. There are odds and ends to work out, most of them tangled and difficult. But in recent years the progress has been so notable that we might well substitute the word "puzzle" for "mystery," and lay aside the traditional tools of disputation in favor of the keener instruments of biochemistry and microbiology.

Judged by current standards, the early philosophical and theological contemplators of the origin of life were not very successful in convincing themselves or others. A century ago, using biblical argument and metaphysical vocabularies, they fought wordy battles against the biologists, who resorted to observations—to those lances that so easily penetrate the fondly held theological shields. The biologists in fact changed the problem from the misty mystery of the origin of self-conscious man to the clean-cut general question of the origin and nature of all life, of monkeys and mice, of algae and oaks, of everything that crawls, flies, swims, breathes, and metabolizes. As supernaturalism retreated a bit, many scientists overcorrected, unfortunately, and embraced a sterile God-excluding mechanistic philosophy, with the result that the sniping goes on although the major action is concluded and a truce is established.

Stars and the Mystery of the Self-replicating Macro-molecule

The astronomer in the past has not been much implicated in the origin-of-life puzzles. He had his hands full, his eyes and mind full, with inanimate origins. How came the comets? How began the rotation of the galaxy? Why are the planets spaced as they are? How started the universe? When, where, why? Mysteries aplenty, and except for some rather silly speculations about the Martians the astronomer had no concern for the apparently remote problems of biogenesis.

But a change has come about. Many sciences have become involved. The origin and age of the planet, and especially its early history, have become critical factors in the problem of the origin of organisms. The astronomer, it appears, does now have something to contribute. The geologist, by way of paleontology, has always been in congenial touch with those who ask about life and its early career on this planet. The weatherman—especially the paleoclimatologist—has his fingers and imagination in the business. And the physicists, chemists, and mathematicians—they are at the bottom of everything that is material, dynamic, electrical.

The Strange Case of the Self-duplicating Molecule requires for a solution, or a substantial approach thereto, the skills of all these specialized detectives, of practically everybody, except perhaps the dogmatic theologian!

But that implication is not quite fair. There are many theologians, respectful of their doctrine, who profit from the cosmic messages of the sciences; they choose to go along constructively rather than retreat awkwardly before the evolving evidence. And the scientists gain from their company; by them the seeming harshness of cosmic law is softened.

Some theologies are not frozen, not fossilized at a given

epoch; their spokesmen recognize the bearing of the advance of knowledge on the tenability of the ancient positions. Some philosophers, not too many, re-examine, re-evaluate, and go forward. By them the cosmologies are reformed to agree with verified data of biology and physics. Moreover, this evolution of doctrine need not be reluctant, gradual, slow. In situations under human control (like man's own reasoning), beneficent mutations should be welcomed and if possible incited. For change, growth, evolution in this live dynamic universe is inherent and wide-spread. "If ought were constant in this world/ Thy turn had never come to thee."

Evolution affects not only stars, galaxies, and planetary crusts, animals, plants, and societies, but also touches social policies, the ethical systems of man, and the religions he fosters. May not science, broadly taken, be the fundamental cultural soil in which we plant and vitalize our religions? Need so many of them remain dated and nonrational?

It seems to me appropriate at this point, before we consider extraterrestrial chemistry in useful detail, to quote a great churchman's views on the verity of mutability and evolution throughout the universe. The excerpts are from an address given in 1951 by the late Pope Pius XII to the Vatican Academy of Science.

At first sight it is rightly a source of wonderment to recognize how the knowledge of the fact of mutability has gained ever greater ground, both in the macrocosm and in the microcosm, according as science has made new progress, as though confirming with new proofs the theory of Heraclitus: "Everything is in flux": *panta rhei.* As is known, our own everyday experience brings to light an immense number of transformations in the world around us, both near and far away, particularly the local movement of bodies. . . . Going still farther, natural science has made known that this chemico-physical mutability is not, as the ancients thought, restricted to terrestrial bodies, but even extends to all the bodies of our solar system and of the great universe, which the telescope, and still more the spectroscope, have demonstrated to be composed of the same kind of atoms. . . .

Nevertheless, in the face of the undeniable mutability of even inanimate nature, there still rises the enigma of the unexplored microcosm. It seemed, in fact, that, unlike the organic world, inorganic matter was in a certain sense immutable. Its tiniest parts, the chemical atoms, were indeed capable of combining in most diversified manners, but they appeared to be endowed with a privilege of eternal stability and indestructibility, since they emerged unchanged from every chemical synthesis and analysis. A hundred years ago, the elementary particles were still regarded as simple, indivisible, and indestructible. The same idea prevailed regarding the material energy and forces of the cosmos, especially on the basis of the fundamental laws of the conservation of mass and energy. . . . The growing knowledge of the periodic system of chemical elements, the discovery of the corpuscular radiations of radioactive elements, along with many other similar facts, have demonstrated that the microcosm of the chemical atom, with dimensions as small as a ten millionth of a millimeter, is a theater of continuous mutations. . . .

The very first modest attempt to break down the nucleus (of nitrogen) goes back to hardly more than three decades ago, and it is only in recent years that it has been possible, by bringing into play tremendous forces, to produce very numerous processes involving the formation and the breaking down of nuclei. Although this result—which, insofar as it contributes to the cause of peace, is certainly to be inscribed among the glories of the century—represents in a field of practical nuclear physics no more than a preliminary step, nevertheless it provides for our considerations an important conclusion, namely, that atomic nuclei are indeed, by many orders of magnitude, more firm and stable than ordinary chemical compositions, but notwithstanding this they are also, in principle, subject to similar laws of transformation, and hence are mutable.

At the same time it was possible to establish that such processes have the greatest importance in the economy of energy of the fixed stars. In the center of our sun, for example, according to Bethe, and in the midst of a temperature which goes as high as some twenty million degrees, there takes place a chain-reaction, returning upon itself, in which four hydrogen nuclei combine into one nucleus of helium. The energy thus liberated comes to compensate the loss involved in the radiation of the sun itself. . . .

If the scientist turns his attention from the present state of the universe to the future, even the very remote future, he finds himself constrained to recognize, both in the macrocosm and in the microcosm, that the world is growing old. In the course of billions of years, even the apparently inexhaustible quantities of atomic nuclei lose utilizable energy and, so to speak, matter becomes like an extinct and scoriform volcano. And the thought comes spontaneously that if this present cosmos, today so pulsating with rhythm and life, is, as we have seen, insufficient to explain itself, with still less reason, will any such explanation be forthcoming from the cosmos over which, in its own way, the shadow of death will have passed. . . .

If we look back into the past at the time required for this process of the "Expanding Universe," it follows that, from one to ten thousand million years ago, the matter of the spiral nebulae [galaxies] was compressed into a relatively restricted space at the time the cosmic processes had their beginning.

To calculate the age [of the solid crust of the earth from the age] of original radioactive substances, very approximate data are taken from the transformation of the isotope of uranium 238 into an isotope of lead (RaG), or of an isotope of uranium 235 into actinium D (AcD), and of the isotope of thorium 232 into thorium D (ThD). The mass of helium thereby formed can serve as a means of control. This leads to the conclusion that the average age of the oldest minerals is at the most five thousand million years. . . .

The pertinent facts of the natural sciences, to which We have referred, are awaiting still further research and confirmation, and the theories founded on them are in need of further development and proof before they can provide a sure foundation of arguments which, of themselves, are outside the proper sphere of the natural sciences. This notwithstanding, it is worthy of note that modern scholars in these fields regard the idea of the creation of the universe as entirely compatible with their scientific conceptions and that they are even led spontaneously to this conclusion by their scientific research. Just a few decades ago, any such "hypothesis" was rejected as entirely irreconcilable with the present state of science.

Cosmo-chemistry and the Astronomer as Detective

We return from these remarks on nuclear physics, the age of the earth, and related subjects to our observation that many scientific disciplines are involved in the technical work on the so-called mystery of life. The astronomer enters the operation in two ways. First, he can contribute information and rational speculation on the age of the terrestrial rocks and the probable temperature conditions on the earth's surface in earlier millennia. His second contribution develops from the attempt to look beyond this planet's organisms and inquire about the spread of biochemical evolution elsewhere. We have already argued the case for the existence of myriads of suitable planets throughout metagalactic space. The question now is what is the likelihood that habitable planets are in fact biologically occupied. We are not specifying what the nature of the life may be, whether closely resembling some terrestrial forms, or basically different. The difference might arise from oxygen scarcity, hydrogen overrichness, atmospheric density too low or too high, or from other conditions that would guide biochemical evolution in ways others than those terrestrial.

It is amazing what grand thoughts and great speculations we can logically develop on this planet—thoughts about the chemistry of the *whole* universe—when we have such a tiny sample here at hand. We can analyze chemically only the samples that we pick up. The earth is composed of six thousand quintillion tons of water, air, and rock (chiefly rock). This quantity is, however, trifling compared with the contents of the sun, which has three hundred and thirty thousand times as much matter; and the sun is but one of more than a hundred quintillion stars. Moreover, we can actually examine chemically only a small fraction of the earth's stony crust. We know the oceans pretty well and the lower atmosphere, but what a minute sample that is for the grand extrapolations.

We do have, to be sure, a little nonterrestrial matter—the incoming meteorites. Currently they are not important additions to the earth's mass and they tell us very little that we do not already know. Through chemical analysis they do show that there is probably nothing new under or near the sun, no chemical elements that we do not find in the solar atmosphere or in the earth's crust. Some of the minerals in meteorites are in combinations different from those we find on earth, but of course we have not been deep into the earth where pressures are high and where the relative proportions of elements differ from those near the surface; the large meteorites are indeed fragments that may have come from the interior of a planet long ago fractured.

Cosmic-ray nucleons from outer space, chiefly protons, bombard our upper atmosphere, and some get all the way through; and of course mass-carrying starlight and some radio waves reach the earth's surface from outer space. But only the meteorites bring outside information of a sort suitable for laboratory analysis. No samples from Mars or Jupiter are at hand, and as yet no rock from the nearby moon. As matters stand, it looks like a severe and nearly complete isolation of inquiring man from the rest of the universe about which he is so curious. Certainly the isolation is complete for the moon-seeking moth and for the sun-fed plants, and the animals that we like to call lower than man—lower in intellect if not in curiosity. They are earth-bound, but man struggles against isolation.

The rather hopeless separation of Homo sapiens from stars and other planets had indeed prevailed until about a century ago when chemistry went to the heavens and man was suddenly in material contact with sun, stars, and glowing nebulosities. No longer need he fret about his isolation. Interstellar distances were no longer a deterrent to knowledge of the chemistries of the wide-spread cosmos.

This is how the wonder came about, and we start with some very ancient human history and proceed to some rather technical concepts that can scarcely be avoided if we are to grasp very important conclusions about the nature of the universe and man.

Rainbows and Stellar Spectra

The rainbow with its spread of colors and its association with storm clouds and the mists of waterfalls and fountains has been a beautiful mystifying spectacle throughout the ages.

The primitives of Nias, for example, "tremble at the sight of a rainbow, because they think it is a net spread by a powerful spirit to catch their shadows."* To the ancient Hebrews it stood as a pledge against the recurrence of catastrophic world-wide floods.† But it has a deeper meaning, a more certain significance to Cosmography; it is a clue to the composite nature of light—a clue that was not followed up until recent centuries.

The refraction of light by water drops, the amount of the deviation depending on the wave length of the light, was recognized as the source of the rainbow's colors long before astronomers made use of this principle in the analysis of starlight. But the complete and correct theory, which explains that the solar spectrum is shown in the rainbow, was not rapidly or easily attained. Aristotle and Seneca missed; Grosseteste of Oxford and Witelo the Silesian in the thirteenth century hit the right theme—raindrop refraction. Later came Theodoric of Freiburg with the interpretation of different combinations of refraction and reflection for the primary and secondary rainbows. He was followed by many improvers of the complicated theory, including big-name scientists like Descartes, Edmund Halley, Newton, Thomas Young, and Astronomer Royal Sir George B. Airy who in 1838 left little unexplained. The rainbow, however, although incited by the sun, is an earth-bound meteorological phenomenon, and our current interest is in starlight as a guide to cosmic chemistry.

In 1666 Sir Isaac Newton's experiments, with prisms of glass and with sunlight that had passed through a narrow aperture into a dark compartment, demonstrated that sunlight is composed of all colors, each color having its own refrangibility (bending). The sun's spectrum was not recognized in all its power, however, until Wollaston of London and the

---

* Sir James George Frazer, *Taboo and the Perils of the Soul* (London, 1936), p. 79.

† Genesis 9:13.

German scientists Fraunhofer and Kirchhoff detected and interpreted the interruptions in the continuous spectrum—those "breaks between the colors." The breaks soon were recognized as dark lines, then as absorption lines, then as clues to the kinds of chemical atoms in the sun's atmosphere. By these scientists and their followers the fertile new Age of Spectroscopy was opened a century ago. Laboratory analysis of light showed that each substance, such as sodium, sulphur, calcium, and iron, when properly heated and excited, had a characteristic pattern of bright radiations of specific wave lengths, specific positions along the spectrum.

In the solar spectrum these patterned radiations become characteristic dark absorption lines. That is, the various kinds of atoms in the sun's outer atmosphere block (absorb) specific wave lengths of the sun's whole-spectrum radiation. Atoms of an element radiate and absorb in the same wave lengths; the continuous background-radiation they absorb is produced by the hot body of the sun.

Notwithstanding the growing evidence yielded by the early spectroscopes, that the lines in the solar spectrum were indicators of the chemical composition of the sun's atmosphere, a famous European scholar, Auguste Comte, declared a century and a quarter ago that we should never be able to know the chemistry of the stars. Only a few decades later Sir William Huggins in England, Father Secchi in Rome, and others were reporting on the chemical constitution of the brighter stars; and with the arrival in the astronomer's tool-box of the photographic plate as an accurate recorder of starlight, the classification of stars, bright and faint, on the basis of their chemistry became big business. At the beginning of this century the work largely centered at the Harvard Observatory in the hands of Antonia Maury and Annie J. Cannon, with important special contributions coming from the Lick Observatory in California, from Potsdam in Germany, from Poulkova in Russia, and the Vatican Observatory. Philosopher Comte notwithstanding, the barriers to extraterrestrial chemistry fell away. This remarkable outcome well illustrates the adage that it is not good for the reputation of one's judgment and imagination to say and believe that anything cannot *ever* be done.

Everywhere: Same Chemistry, Same Physics

A few statements on the chemistry of stars as shown by spectrum analysis will assist in our orientation among the atoms. To those who find the language rather technical, I suggest that this section be skipped. The last sentence of the chapter states the major conclusion.

(1) Only the surface of a star, of course, can be directly studied through the analysis of light, but astrophysical theory and mathematical "boring tools" boldly plunge beneath the surface and lead us to report confidently on the chemical composition that must exist far down under to account, first, for the kind of spectrum we find at the surface and, second, for the total brightness of the star.

(2) Miss Cannon's classification puts the stars into sixty different classes and subclasses, arranged in order of surface temperature. The temperature prescribes the appearance or absence of the various spectral lines, and also determines the color. Red Betelgeuse, a cool giant star, has a spectrum rich in absorption lines; yellowish Polaris has many fewer, and bluish Rigel in Orion is so hot at the surface that only the absorption lines of hydrogen and helium are prominent. More recent work in America and Sweden has increased the number of subclasses by adding to Miss Cannon's temperature-based classification a subclassification that depends on the intrinsic luminosity (candle power) of a star, thus providing in the description of stellar spectra a sort of second dimension. This subclassification is very useful in estimating the candle powers and subsequently the distances of many types of stars.

(3) About twenty per cent of the stars, according to the large sampling in Miss Cannon's catalogue, are similar to our sun in spectrum, and therefore more than forty thousand neighboring stars are sunlike in color and surface chemistry. This proportion probably holds on the average throughout our galaxy and others, but with a higher percentage of solar stars in the nuclei than in the spiral arms.

(4) The spectra of whole galaxies are difficult to photograph clearly. So far as we have studied them, however, they are what we would expect—a composite of all spectral classes. They are, in fact, much like the sun's spectrum, Class G0, which is in the middle of the spectral series from hottest Class B to coolest Class M.

(5) Some of the spectra of individual stars in remote galaxies are now known; they are completely like those of stars in our galaxy—an observation that again emphasizes a common chemistry throughout the universe we explore. In the nearest of the external galaxies, the Star Clouds of Magellan, the spectra of hundreds of stars have been photographed. All are of familiar types. All the many classes of stars and nebulosities found in the Magellanic Clouds have their counterparts with us.

(6) In these two nearest of galaxies, which are irregular in form, being neither symmetrically spiral nor smoothly spheroidal, are found nearly all the familiar kinds of stars that are variable in light—those natural indicators of evolutionary processes. Their spectral classes range from reddish M and N (long-period variables), through yellowish K, G, and F (classical cepheids), to A and B at the hot end of the spectrum.

(7) Among the variables are eclipsing binaries, mostly of spectral class B. Such doubles can be analyzed through studies of their light variations, much as we work out the characteristics of the famous eclipsing star Algol of our galaxy. Their masses, relative motions, temperatures, densities, sizes can be calculated. We find that the same celestial mechanics, the

same gravitational and radiational laws, in brief, that the same physics prevails in these other galaxies as here at home. We are led to the important generalization that we have a common physics as well as a common chemistry throughout the whole of the explorable universe.

(8) Finally, in these comments on the knowledge that the multi-colored rainbow has led us to, attention is called to another ancient observation, which now baffles us much less than it baffled our remote ancestors. This newer clarity has also come through spectrum analysis. I refer to the Sword of Orion, and its hazy central star. The spectroscope has been able to "loose the bands of Orion" by showing what Job could not know, namely, that the Orion Nebula is a mass of oxygen, hydrogen, carbon, and nitrogen gas, which is excited into radiation by the superhot stars of its vicinity.

(9) The spectroscope has, in fact, revealed the wide distribution in interstellar space of the gaseous stuff that stars are made of; and, of higher significance at this point in our exposition, it shows that such shining nebulous material exists also in the Magellanic Clouds and the great Andromeda Galaxy, and indeed throughout the Metagalaxy. This is an important revelation in that it indicates that the universe is not yet completed, and may never be. It suggests that probably galaxies as well as stars (and perhaps planets) are being formed currently in response to the natural laws of gases, celestial motions, and light propagation—laws that we already understand pretty well.

Radio Signals from Outer Space

As a postscript to this chapter on orientation by way of the spectrum, it is appropriate to mention that the recently developed radio telescopes, working in wave lengths from a centimeter to thirty meters and longer, have entered the field miraculously. (I use that word loosely, since miracles connote the supernatural which is here hardly an acceptable concept.) For example, the large radio-wave receivers find in all parts of the Metagalaxy, from the nearby spiral arms of our own system out to clusters of galaxies a hundred million light years distant, the weak radiation from cold interstellar and intergalactic hydrogen atoms. These radiation waves, eight inches from crest to crest, were first predicted from atomic theory, then detected in our Milky Way, and then in a rush exploited by radio astronomers and electronic engineers in Australia, Holland, and America. Here is a listing of some of the recent discoveries of radio astronomy, all of which have cosmographic relevance.

(1) The Magellanic Clouds are found to be embedded in a low-density medium of neutral hydrogen—that is, neutral in electric charge, in contrast to ionized hydrogen such as that which helps to illumine the Orion Nebula.

(2) The spiral arms of our own galaxy have been traced by discovering the excess of neutral hydrogen within them.

(3) The expansion of the universe is confirmed through finding a consistent red shift in radio waves that are half a million times longer than the visual light waves where first the red shift was discovered and measured.

(4) The existence in interstellar space of drifting dust, which is one component in the composition of evolving stars, is confirmed through the radar detection in the earth's atmosphere of innumerable meteors.

(5) Natural radio "signals" are now received from some of the planets, notably from Jupiter's cloudy surface. Electric storms in and around Jupiter's so-called red spot are a suspected source of the signals. Do we have here the thunderbolts of Jove?

(6) The sun, the Milky Way, the wrecks of supernovae, and hundreds of unidentified "radio stars" in remote space all contribute signals in the measurable radio frequencies, making

this new phase of astronomical science sensational, if not miraculous.

Is it necessary to add that the use of the word "signal" does not imply a belief that any of these impulses originate with a living being? Radio signals need not always require biological agents. Lightning provides a signal, as do whirling sunspots, and they have natural physical causes.

To summarize the last two sections in a sentence: Spectroscopes, radio telescopes, scientific methodology, and the mathematics of the physical scientist have joined in revealing a uniformity of structure, composition, and behavior throughout the cosmos and guarantee to the searcher on Planet No. 3 that he can reasonably assume that what holds here holds widely.

★ 7 ★

The Fourth Adjustment

In the past history of the evolving human mind, with its increasing knowledge of the surrounding world, there must have been a time when the philosophers of the early tribes began to realize that the world is not simply anthropocentric, centered on man himself. As society developed, the village attained central significance—a natural view supported by the evidence of a circular horizon and by the increasing vagueness of the world as one increased the distance from home. But the higher civilizations of the Near and Middle East (and perhaps elsewhere) became increasingly conscious, a few thousand years ago, of the daily revolving sun, stars, and wandering planets. The navigators detected evidence of the curvature of the surface of the oceans and of the earth. The sphericity indicated thereby led to the belief that the center of the earth rather than a surface locality was the center of the visible universe. This view was considered to be consistent with the apparent motions of moon, planets, sun, and stars. The *geocentric* concept thus became the common doctrine in many of the most civilized nations.

This first adjustment of man to the rest of the total material universe was only mildly disturbing to his ego, for man appeared, on pretty good evidence, to surpass all other living

NOTE: This chapter is adapted from an article that appeared in *The American Scholar*, Autumn, 1956.

forms. He saw little reason to be humble. He personally was not central, but his earth had that distinction.

From Geocentric to Heliocentric

The second adjustment was the abandonment of this earth-center theory. The new hypothesis was not generally accept-able in the Western world until the Copernican Revolution of the sixteenth century soundly established the heliocentric concept. The liberal philosophers and eventually the church fathers yielded to the scientists' theory of a universe centered on our sun. It was a slow shift, for man is a stubborn adher-ent to official dogma. In time, however, he accepted the sun as the center not only of the local family of planets but also of the total sidereal assemblage; and he long held that view. But it, too, was a fallacy. Another shift was in the making as soon as the sun was recognized as an ordinary star; but only when modern telescopes reported on globular star clusters, galaxies, and cepheid variables did a further adjustment be-come imperative.

The earth-centered cosmology had been given up in favor of the sun-centered system very reluctantly. And likewise, later, in spite of increasing evidence requiring a further change, the scientists, philosophers, and laymen held dogged-ly to the heliocentric view. Was this holding fast because of vanity—because of the feeling, cultivated by the unscientific dogmatists, that man is of paramount significance in the world of stars and space-time?

From Heliocentric to Sagittarius and Beyond

There are several better reasons for this second erroneous concept—the heliocentric theory; they are quasi-scientific explanations. For example, the Milky Way follows a great circle; it is a band of light that divides the sky into two practically equal parts. Also, it is of about the same brightness in all parts. By implication, therefore, the sun and earth are centrally located. A second evidence is that the numbers of stars seemed to the early census-takers to fall off with distance from the sun as though it were central; and such a position for his star among the stellar millions brought to man a dignity of position not at all disagreeable. But again it was an illusion.

As late as 1917 the leaders in astronomical interpretation held that the sun was central, or at least very near the center of the sidereal universe. (The galaxies were then not recognized officially as other great stellar systems.) The introduction of the period-luminosity relation for cepheid variable stars as a sounding tool and the determination of the distances and distribution in space of the globular star clusters first indicated the eccentric position of the earth, sun, and surrounding stars in the flattened stellar system which is made manifest by the star-crowded Milky Way.

Gradually came other evidence that the billion-starred nucleus of our spiral galaxy is remotely distant through the southern constellations of Sagittarius, Ophiuchus, and Scorpio. With that thrust into the stellar depths, the heliocentric theory of the stellar universe struggled briefly, weakened, and died.

The center of the galaxy is not near at hand among the bright stars that define those southern constellations, for they are but a few hundred light years away. The center of our galaxy, we have found, is more than twenty-five thousand light years distant. The billions of stars in that nucleus together make the large white glow in the southern Milky Way which we call the Sagittarius star cloud.

The shift from the geocentric to the heliocentric concept doubtless had some philosophical impact in the sixteenth century, but not much. After all, the hot, turbulent, gaseous sun is no place for the delicate array of biological forms in which man finds himself at or near the top. Earth-center or sun-center seemed to make little difference to cosmic thinking. From the deathbed of Copernicus to the birth of this century and later the prevailing heliocentric concept of the stellar universe incited little if any philosophical uneasiness.

But then, with the rapidly increasing accumulation of astronomical information, came the inescapable need for this third adjustment—one that should have deeply affected and to some extent has disturbed man's concern about his place, his career, and his cosmic importance.

This shift of the sun and earth to the edge of our galaxy has considerably eroded human pride and self-assurance; it has carried with it the revelation of the appalling number of comparable galaxies. We could accept rather cheerfully the Darwinian evidence and argument of our animal origin (although the theologians of a century ago found it strong medicine), for that evidence still left us, we believed, at the summit of all terrestrial organisms. But the abandonment of the heliocentric universe, on the basis of dependable astronomical evidence, was certainly deflationary from the standpoint of man's position in the material world, however flattering such advances of human knowledge were to the human mind.

The galactocentric hypothesis puts the earth and its life on the outer fringe of one galaxy in a universe of millions of galaxies. Man becomes peripheral among the billions of stars of his own Milky Way; and according to the revelations of paleontology and geochemistry he is also exposed as a recent, and perhaps an ephemeral manifestation in the unrolling of cosmic time.

At this point we pause for a somber or happy thought, one that is somber or happy depending on one's mood. With the advance of science, and with the retreat of superstition and of belief in the supernatural, we have in recent centuries gone so far and so firmly in our orientation of man in the universe that there is now no retreat! The inquiring human has passed the point of no return. We cannot restore geocentrism or even heliocentrism.

The apes, eagles, and honey bees, with their specialized skills and wisdoms, may be wholly content to be only peripheral ephemerals, and thus miss the great vision that opens before us. For them egocentrism or lococentrism may suffice; for us, no! And since we cannot (and will not) go back to the cramped but comfortable past without sacrificing completely our cultures and civilizations, we go forward; and then we find that there is another chapter in the story of orientation.

Biological Orientation

Another shift must be made, for we are concerned in this discussion not only with the location of our earth in the time and space of the physical world, but with our own location in the biological world. The downgrading of the earth and sun and the elevation of the galaxies is not the end of this progress of scientific pilgrims plodding through philosophic fields. As intimated on previous pages, the need for the further jolting adjustment that now arises above the mental horizon is neither wholly unexpected by workers in scientific fields, nor wholly the result of one or two scientific discoveries. It is a product of the age. We turn from astronomy to the overlap of a dozen other sciences and ask about the spread of life throughout the universe.

As unsolicited spokesmen for all the earthly organisms of land, sea, and air, again we ask the piquant question: "In this universe of stars, space, and time, *are we alone?*"

From among the many thoughts and measures that promote this Fourth Adjustment of Homo sapiens in the galaxy of galaxies, three phenomena stand out as most meriting our further consideration. The first refers to the number of stars, the second to the catastrophes of ancient days, and the third to the origin of self-replicating molecules. They are worth brief summarizing at this point although the first two to some extent have been presented in earlier chapters, and the third will be the main theme of Chapter 9.

To the ancients only a few thousand stars were known; to the early telescopes, however, a million; and that astounding number has increased spectacularly with every telescopic advance. Finally, with the discovery that the so-called extragalactic nebulae are in reality galaxies, each with its hundreds or even thousands of millions of stars, and with the inability to "touch metagalactic bottom" with the greatest telescopes, we are led (as shown in Chapter 5) to accept the existence of more than 10^{20} stars in our explorable universe, perhaps many more.

The significance of this discovery, or rather of this uncovering, is that we have at hand—that is, the universe has at hand —more than one hundred million million million sources of light and warmth for whatever planets accompany these radiant stars.

(The number of stars and their ages are of course not humanly comprehensible in the usual terms—too many stars, too much space, too many years for minds that are accustomed to operate in serially countable numbers. The macrocosmos transcends our counting. And comprehension is not simplified when we turn to the atomic *micro*cosmos and point out that in our next breath we shall each inhale more than a thousand million million million atoms [10^{21}] of oxygen, nitrogen, and argon.)

The second phenomenon, the expanding Metagalaxy, bears on the question: Do planets accompany at least some of the stars that radiate energy suitable for the complex biological activity that we call life?—a question that was asked and tentatively answered in Chapter 4.

We now accept the strong observational evidence of a universal redward shift in the light received from distant external galaxies, and accept also the interpretation of that red-shift as a result of the systematic scattering and diffusion of galaxies and the expansion of the universe. The speed of the mutual recessions is about thirty-five miles a second for galaxies separated by a million light years; twice as fast for galaxies at twice the distance apart; three times at thrice the distance, and so on. The exact numerical values are still under investigation, as is the possible failure to maintain at great separations this uniform increase of scattering speed with distance.

The Turbulence of Long Ago

The rapid dissipation of the Metagalaxy in all directions naturally turns thought to the situation of a year ago when the galaxies were closer together, and to a century, a millennium, a billion years ago. There was of course, as we go back in time, an increasingly greater concentration of the new spreading cosmic units (galaxies). The average density of matter in space at present is very low—something like 10^{-30} grams per cubic centimeter, which on terrestrial standards is a veritable super-super vacuum. A few thousand million years ago, however, the average density in the unexpanded universe must have been so great that collisions of stars and gravitational disruptions of both planets and stars were inevitably frequent.

Now here is an important coincidence. The crust of the earth, radioactively measured, is also a few thousand million years old. Therefore the earth and the other planets of this planetary system were born in those crowded days of turbulence and disastrous encounters.

At that time countless millions of other planetary systems must have developed, for our sun is of a very common stellar type. And stars of nonsolar types must have also participated in the cosmic turmoil. (Our sun, a primitive compared with many blue and red giant stars of recent origin, is so common that in Miss Cannon's famous spectrum catalogue we find some forty thousand sunlike stars, all in our immediate neighborhood.)

Other ways in which planets may be formed, other than this slam-bang process of the early days, have been proposed by astronomers and other scientists (Chapter 4). For example, the contraction of protostars out of the hypothecated primeval gas, giving birth to protoplanets on the way, is an evolutionary process now widely favored. It would imply the existence of countless planets.

The head-on collision theory of planetary origin also has been favorably considered in various versions. But the stars are now so widely dispersed that collisions must be exceedingly rare—so very unlikely, in fact, that we might claim uniqueness throughout all creation for ourselves, if planet birth depended only on such collisional procedure. But that vanity cannot be easily maintained, since the expanding universe discovery has shown the crowded, collision-filled conditions when our earth emerged out of the chaos.

Passing over details, we again state the relevant conclusion: *Millions of planetary systems must exist*, and billions is the better word. Whatever the methods of origin, and doubtless more than one type of genesis has operated, planets may be the common heritage of all stars except those so situated that planetary materials would be swallowed up by greater masses or cast off through gravitational action. In passing, we recall that astrophysics has shown that our kind of chemistry and physics prevails throughout the explorable universe. There is nothing uncommon and special here or now.

Remembering our 10^{20} stars and the high probability of millions of planets with suitable chemistry, dimensions, and distance from their nutrient stars, we are ready for the question:

On some of these planets is there actually life? Or is that biochemical operation strangely limited to our planet, limited to No. 3 in the family of the sun, which is an average star located in the outer part of a galaxy that contains a hundred thousand million other stars—and this local galaxy but one of millions of galaxies already on the records?

Is life thus restricted? Of course not. We are not alone. And we can accept life's wide dispersion still more confidently when our third phenomenon is indicated.

To summarize in four sentences what we shall spell out in some detail in Chapter 9: Biochemistry and microbiology, with the assistance of geophysics, astronomy, and other sciences, have gone so far in bridging the gap between the inanimate and the living that we can no longer doubt but that whenever the physics, chemistry, and climates are right on a planet's surface, life will emerge, persist, and evolve. The mystery of life is vanishing. Objective science is replacing the subjective miraculous. The many researches of the past few years in the field of macromolecules and microorganisms have now made it quite unnecessary to postulate miracles and the supernatural for the origin of life.

The step in human orientation that I call the Fourth Adjustment is ready for the taking, if we care to accept that opportunity. The scattering of galaxies, the abundance of stars, and the structure and habits of macromolecules on warm, moist, star-lit planetary surfaces have prompted this further and most important adjustment in the understanding of the place and functioning of life in the universe. The acceptance of the evidence and the belief that the biological development on this planet is not unique and that varied and highly elaborated sentient life is abundant and widely distributed, have led to the most important step of all in the orientation of Homo in the material world.

Have we come now to the end of the journey, or are there

other steps ahead? In view of the rapid growth of scientific techniques and the continual exercise of the logical imagination, it would not be wise to suggest that we shall never *never* find need for further adjustment of the concept of man's place in the universe—that we shall never discover a reason for an orienting adjustment that transcends both the physical and biological orientations, which are now represented respectively by the third and fourth adjustments.

A fifth adjustment might be in the psychological realm, or in the "negative matter" world, or in one of those fanciful existences where our Metagalaxy is only an atom in some superuniverse, or in the equally droll (and equally possible) existence where our electrons are the galaxies in some microcosmic universe that is below our measures and our knowing.

⋆ **8** ⋆

A Digression on Great Moments

Cosmic energy in the form of sunshine and the earth's leaking body heat, in cooperation with the organic compounds of the primeval "soup" in the earth's primitive shallow seas, resulted long ago in the evolution of green leaves and stems, and the eventual development of the higher plants and animals. The early and continuing biochemical reaction called photosynthesis was paramount in this cooperation. Here we point to the beginning of the collaboration of radiant energy and organic molecules as a decisive step in the life episode on the terrestrial surface. It was the start of Operation Chlorophyll, without which we would not be. So far as life is concerned, it was indeed one of the Great Moments of the universe; if it could happen here it could happen elsewhere, with perhaps equally momentous consequences.

The idea of a particular Great Moment, a critical turning point in the manifold evolution of the material universe, is an intriguing thought. We have college courses on Great Ideas, reading lists of Great Books, and books on Great Men. There are Great Expectations, High Lights of the Year, and so on. Why not look for the Great Epochs in Cosmography?

Before the primitive plants got established the earth's rocky crust was essentially barren biologically, as the moon is now;

NOTE: This chapter is adapted from an article that appeared in *The American Scholar*, Summer, 1957.

and then began photosynthesis, an operation for which we shall always be grateful. What are some other equally revolutionary episodes? They may be for the most part too hidden for us to detect, for we are indeed considerably impotent and ignorant in the face of Nature's complexities. But some of them are discernible.

Different choosers of epochal events would probably make different choices. I suggest the following:

(1) The explosion of the all-including Primeval Atom (if there was one)—a violence that according to current hypotheses has resulted successively in the expanding universe, the birth of the chemical atoms, and the formation of galaxies and stars; also in the origin of planets, some of which, with moist and rocky surfaces, were suited to the emergence of varied organisms, among which are those that speculate on Great Moments.

(2) The fortunate "cooperation of charm and countercharm" (as an ancient Chinese saying puts it) that permitted natural laws to provide for the existence of the cohesive atomic nucleus. That concept is a rather hard one—it takes a bit of grasping. I am reminded that a famous physicist who explores the nature of atomic nuclei once confided that it is only by a narrow squeak that matter exists at all.

But perhaps this ubiquitous nuclear coherence does not involve the one-directional passage of time and should be listed with Momentous Facts rather than with factual moments.

(3) The lightning strokes, or other natural energy manifestations, that helped to synthesize out of the preprimeval atmosphere of the earth (and comparable planets), where methane and ammonia, water vapor and hydrogen prevailed, the amino acids that underlie the proteins that underlie organisms. This means, in short: Nature's synthesis of the first continuing life—or at least the vital first step toward the origin, on the earth or elsewhere, of "material organizations perpetuating their organization."

(4) The afore-mentioned invention or "accident" of photosynthesis, including its becoming a biological habit. This complicated device brought life's energy down from the sun.

(5) The issuance from the shallow waters of seashore, lake, and river of primitive animals that learned to take their oxygen raw—an occurrence of some four hundred million years ago for this planet, perhaps much later for many others. Some of these primitives that crawled or flopped ashore were the forerunners of insects; some were the ancestral frogs and other early amphibia. Below the surface of the water the primitive plants and animals got their oxygen, which burns and builds, in diluted dosage. They found it captured by waves or ripples from the overlying air. On land the inhaling of concentrated oxygen was an early accomplishment of major moment.

The specialized breathing apparatus developed variously. For instance, we higher mammals long ago abandoned our ancestral gills and developed in their stead a bellows arrangement (lungs) equipped with strings (vocal chords) which permits conversation and song in addition to the traffic in oxygen and carbon dioxide. In other words, a happy by-product of the development of an efficient apparatus for inhaling needed oxygen and exhaling undesired carbon dioxide was the production of an ingenious vocal means of communication. For the transfer of information to each other it made us not dependent, as are the ants, on the twiddling of antennae, or dependent, as are the bees, on the waggling of our abdomens.

(6) The "invention," that is, the rapid mutational development, of insect wings. These appendages, which have made the mighty class of Insecta so dominant (a million kinds, a trillion individuals), and in many ways so important in the life of the planet, did not arise through the slow adaptation of existing appendages to a new use. The wings of flying mammals (bats) and of flying neoreptiles (the birds) have arisen through the adaptation of the forelegs. The insect wings were acquired without the sacrifice of standard appendages. They

are something entirely new and extra, and serve in part as an escape mechanism. The survival since the Paleozoic Era of many orders of insects was doubtless dependent on their overcoming some of the precariousness of life on the ground where they could not easily escape their devourers by running, crawling, or hiding. These primitive insects that could take to the air on occasion, long before there were birds to exploit the art of traveling in a gaseous medium—those early winged insects met one important test for the survival of the species; they got away! Their escape mechanism was indeed novel and well designed.

We could list additional momentous items, such as the descent of our ancestral Hominidae from the trees, the discovery of the use of fire, the announcement of the laws of motion and of gravitation. But some of them, like insect wings, may be too local to be considered as Great Moments for the universe. Most of those numbered above obviously are gradual changes, not sudden unique inventions. In fact, it might be better to use the word "epoch" or "era" rather than "moment."

Incidentally, if partial comprehension and description of the cosmos is in itself a part of Cosmography and is an important item in the whole cosmic scheme, then the big mutations in the primate cortex are master moments, and so was the discovery of the wheel, and of the calculus, and the subconscious. But let's skip this anthropocentricity and get back to the stars and the emergence of organisms.

★ 9 ★

Toward the Emergence of Organisms

~~~~~~

It is not enough simply to assert the apparent inevitability of the emergence of life on a planet's surface when conditions are right—when the chemicals proper for protoplasm are available, the mass, temperature and motions of the planet are suitable, and the weather is good. To say that biochemical evolution is natural and insistent is not the whole story. We should report further on how Nature builds up self-replicating molecular aggregates and detail some of the steps in this process that is the essence of the phenomenon called living.

The subject of life's origination has a sultry history. Incompetent observations and a willingness to make silly deductions kept alive for centuries the thesis that low forms of life could spontaneously arise from filth. Pasteur helped kill that heresy. He and his fellow scientists for a time were opposed by a few scientific critics. But the worst blocking of the scientific advance against mystery in the origin of life came from the religious fortifications, which were manfully defended by cloth and laity. It was stoutly held that although man could make alloys in his laboratories that Nature had not made, and could develop new hybrids of plants and animals, and create societies, he was, is, and always will be impotent in the laboratory production of the biologically alive. He could destroy life, visible and invisible, but the original creation of life was held to be exclusively in the inscrutable hands of the Almighty.

To seek the origin of life was to peer into forbidden places; it approached the blasphemous.

Thus cramped by dogma, science had to enter into research on human biogenesis indirectly, by way of studies of the simplest organisms and the nature of life at its threshold. But the dogmatic opposition ceased to prevail rather long ago. In the past century our knowledge of the universe expanded so profusely in its many phases and dimensions that ancient prophesies could no longer, even with "interpretation," encompass it all. The biochemists and the microbiologists of the twentieth century began to observe, experiment effectively, and speculate on the beginnings of the primitive life which, the sedimentary rocks report, was already here a billion years ago.

Pioneering by Oparin and Haldane

For example, ignoring magic, forswearing superstition, J. B. S. Haldane ventured in 1928 an essay in which he called attention to the probable early conditions, physical and chemical, on the surface of the earth, and their suitability for the natural synthesis of the animate out of the inanimate. The primitive atmosphere of the earth, it is now widely (but not universally) believed, was free or almost free of oxygen in an uncombined gaseous state. Oxygen is a hungry element and takes every opportunity to combine with receptive atoms. In the form of H_2O (water) and SiO_2 (silica) there was always much terrestrial oxygen—in fact, oxygen is one-half of the crust of the earth, nine-tenths of the oceans, lakes, and rivers, and nearly a quarter of the present atmosphere. But in early times the air had little free oxygen. The thousand trillion tons now in the earth's atmosphere has come in considerable part during the past billion years from the "breathing" of vegetation. It is a by-product of photosynthesis, supplemented somewhat by the breaking up of water vapor by short-wave solar radiation in the upper atmosphere.

Water vapor was dominant in the primitive atmosphere, along with carbon dioxide, methane, and nitrogen in the form of ammonia gas (NH_3). Methane and ammonia still appear in the atmosphere of the cold planet Jupiter, although now practically gone from the earth's air. Free nitrogen and free oxygen constitute about ninety-nine per cent of the present atmosphere, for the original prevailing water vapor condensed, as the planet cooled, into oceans, lakes, and wet soils.

The absence from the primitive atmosphere of atomic and molecular oxygen would indicate, of course, that ozone, the triple-atom form of oxygen, was then also absent. Such absence must have been of considerable biogenetic significance, for the *ozone barrier* in our atmosphere, some twenty miles above us, which now so benevolently protects us tender growths from lethal ultraviolet radiation, was probably not present in the early days, two to four billion years ago. Throughout long eons it may have still been thin and penetrable, and became really effective only when photosynthetic vegetation had released oxygen in large amount from the water and carbon dioxide.

The chief importance of the early lack of ozone would be that the short-wave ultraviolet radiation from the sun could then unhindered bring energy of special potency into the shallow waters (whenever water vapor clouds did not obstruct). The penetrating ultraviolet radiation was no doubt one of three or four sources of energy that could participate in the prechlorophyll generation of primitive life. The others were atmospheric electrical discharges (lightning), gamma radiation from decaying radioactive elements, and possibly the issuing body heat of the earth, such as the heat from geysers and volcanoes.

Short-wave radiation, as H. J. Muller and others have shown, also expedites biological mutations. In preozone days the ultraviolet may have greatly speeded and diversified organic evolution. Now it is blocked.

Haldane's pioneer speculation* on early conditions suitable for the emergence of terrestrial life refers to the energy in short-wave radiation:

> Now, when ultraviolet light acts on a mixture of water, carbon dioxide, and ammonia, a vast variety of organic substances are made, including sugars, and apparently some of the materials from which proteins are built up. This fact has been demonstrated in the laboratory by Baly of Liverpool and his colleagues. In this present world such substances, if left alone, decay—that is to say, they are destroyed by microorganisms. But before the origin of life they must have accumulated till the primitive oceans reached the consistency of hot dilute soup. Today an organism must trust to luck, skill, or strength to obtain its food. The first precursors of life found food available in considerable quantities, and had no competitors in the struggle for existence. As the primitive atmosphere contained little or no oxygen, they must have obtained the energy they needed for growth by some process other than oxidation—in fact, by fermentation. For, as Pasteur put it, fermentation is life without oxygen.

Independently of Haldane, the Russian scientist A. I. Oparin was meditating on the origin of life; he also was unhampered by religious preconceptions. Already in 1923 a preliminary booklet was published by him in Russian; thirteen years later his volume *On the Origin of Life* appeared. It has become a classic. The eighteen-page introduction by S. Morgulis to the second edition of the English translation of Oparin's book is in itself an important contribution to the subject. Since 1950, as Oparin's analysis has become more widely known, several

---

* *Science and Life* (London, 1928).

important researches related to biogenesis have been carried out. Studies in photosynthesis and biochemical work on viruses are entering the field. Powerful electron microscopes analyze the macromolecules. In particular, the experimental work of Stanley Miller, in Dr. Harold Urey's laboratory in Chicago University, is noteworthy. He assembled a sample of the assumed primeval terrestrial atmosphere, composed of the gases methane, ammonia, water vapor, and hydrogen, bombarded it with an electrical discharge (simulating the primeval lightnings) and produced amino acids and other organic compounds. These amino acids, as mentioned above, are the all-important constituents of organisms. The work has been repeated and extended in laboratories at Yale University, the Carnegie Institution of Washington, and Oak Ridge. It will be much extended, for through this simple but technically difficult experiment we have made a long step across the ground that separates the unquestionably inanimate from the unquestionably alive. Miller's report on his pioneer synthesis of the molecular bases of life appears on page 122.

The thoughts of Oparin and his translator on the subject of the emergence of life can best be presented in a few trenchant quotations—all of them rich in meaning to anyone who would grope in the dim past for answers to one of the basic questions of cosmogony. Necessarily, technical terms must be used, and the general reader may therefore prefer to skip some of the following quotations.

To begin with, biochemist Oparin emphasizes the fact that the long-enduring assumption that well-organized life first came through supernatural creation, is irrevocably dead. Also the thesis that life was coeval with matter cannot be maintained—and the evolution of inanimate molecular matter cannot be denied. The astrophysicist would add that atoms as well as molecules have also evolved in the past times we explore, and indeed even now are evolving.

One must first of all categorically reject every attempt to renew the old arguments in favor of a sudden and spontaneous generation of life. It must be understood that no matter how minute an organism may be or how elementary it may appear at first glance it is nevertheless infinitely more complex than any simple solution of organic substances. It possesses a definite dynamically stable structural organization which is founded upon a harmonious combination of strictly coordinated chemical reactions. It would be senseless to expect that such an organization could originate accidentally in a more or less brief span of time from simple solutions or infusions.

However, this need not lead us to the conclusion that there is an absolute and fundamental difference between a living organism and lifeless matter. Everyday experience enables one to differentiate living things from their non-living environment. But the numerous attempts to discover some specific "vital energies" resident only in organisms invariably ended in total failure, as the history of biology in the nineteenth and twentieth centuries teaches us.

That being the case, life could not have existed always. The complex combination of manifestations and properties so characteristic of life must have arisen in the process of evolution of matter. A weak attempt has been made . . . to draw a picture of this evolution without losing contact with the ground of scientifically established facts. (A. I. Oparin, *The Origin of Life,* translation and Introduction by S. Morgulis [2nd ed.; New York, Dover Publications, 1953] pp. 246, 247.)

As the Earth Cooled Off

Whichever of the three or more plausible methods of planetary origin actually prevailed for our system, there can be little doubt but that the earth's crust passed through a stage of molten or at least hot rocks, and certainly a stage of hot atmosphere. As a consequence, there was for a time more hydrogen, helium, and other gases of light atomic weight escaping into space than at present; but otherwise the earth's chemical composition involved the same atoms as now. The molecular compounds, however, were evolving as the rocks, oceans, and air were giving up their heat to the cold of interstellar space.

It is beyond doubt that during the [earliest history of the earth] (especially during the early period of the existence of hydrocarbons), the physical conditions on the earth's surface were different than now: the temperature was much higher, the atmosphere had a different composition, light conditions were different, etc., but in this there is nothing unusual or mysterious. Quite the contrary, these conditions are more or less well known to us and we can not only easily picture them to ourselves but we can even reproduce them, to a large extent, in our own laboratories. Nevertheless, they do not furnish an explanation of how life had arisen on our Earth. And it is not difficult to understand this because knowledge of the external physical conditions is not sufficient for the solution of the problem of the origin of life.

It is also necessary to take into consideration the inherent chemical properties of the substances from which, in the last analysis, living creatures were formed. The study of the behavior of those substances under given external influences will indicate the path which the evolution of organic substance has followed. This approach to the problem is justified especially by the fact that only at the beginning of this evolutionary process were the environmental conditions of existence different from those of our own natural environment. From the time when the primary ocean came into being, the environment in which organic substances existed resembled our own so closely that we may safely draw conclusions about the progress of chemical transformations on the

basis of our knowledge of what is happening today.
[Oparin, pp. 105-106.]

Thus it came about, when our planet had cooled off
sufficiently to allow the condensation of aqueous vapor
and the formation of the first envelope of hot water
around the Earth, that this water already contained in
solution organic substances, the molecules of which were
made up of carbon, hydrogen, oxygen and nitrogen.
These organic substances are endowed with tremendous
chemical potentialities, and they entered a variety of
chemical reactions not only with each other but also with
the elements of the water itself. As a consequence of
these reactions, complex, high-molecular organic compounds were produced similar to those which at the
present time compose the organisms of animals and
plants. By this process also the biologically most important compounds, the proteins, must have originated.
[Oparin, p. 248.]

Adventures in the Hot Thin Soup

As noted on earlier pages, an all-satisfying definition of life
is not easily attained. "Molecular organizations perpetuating
their organizations" is a rather lifeless attempt, even if we un-
derscore perpetuating. Morgulis objects to the phrase "Origin
of Life" as suggesting a single or sudden event—a leap from
the inanimate to throbbing life—which most certainly is not
correct; a better phrase, he suggests, for the title of Oparin's
book would be "Life's Coming into Being." But this writer
would boggle at that word "being."

To most people Life connotes something that crawls,
creeps or at least wiggles if not by means of well articu-
lated appendages at any rate by temporary protoplasmic

protrusions, or cilia, or delicate flagella. Life need not perhaps be visualized in the form of a stalking elephant but to the layman it may seem inconceivable except as some unicellular organism of microscopic dimensions. But even the most primitive unicellular organism has a complexity of structure and function that staggers the mind and is removed from the beginnings of life by a genealogy extending for millions upon millions of years. Possibly, as Oparin so convincingly tells us, it all began some two billion years ago as a venture in colloidal systems of microscopic size separating from the "hot thin soup," to use Haldane's happy description of the primordial ocean.

The biologist, unlike the layman, knows no lines of demarcation separating plant life from animal life, nor for that matter living from non-living material, because such differentiations are purely conceptual and do not correspond to reality. [Morgulis, pp. vii-viii of the translator's introduction to Oparin's book.]

The origin of life was not an occurrence ascribable to some definite place and time; it was a gradual process operating upon the Earth over an inconceivably long span of time, a process of unfolding which consumed perhaps more millions of years than was required for the evolution of all the species of living things. It is one of Oparin's great contributions to the theory of the origin of life that he postulated a long chemical evolution as a necessary preamble to the emergence of life. One might think of the evolutionary process passing through three distinct chemical phases, from inorganic chemistry to organic chemistry, and from organic chemistry to biological chemistry. [Morgulis, pp. vi-vii.]

As long as the cell is considered as the unit of life, the origin of life must remain a paradox. But like the erstwhile atom in chemistry, the cell has lost its prestige as the ultimate unit in biology. Both the atomic and the cellular theories have become obsolete. The cell, like the "indivisible" atom, is now recognized as a highly organized and integrated system built up from extremely small and distinct particles. Whether the ultimate particles of life have been found and identified is very doubtful, some of the units themselves being highly organized entities, but the concept of a cell as a unit of life has been thrown out of the window together with the atom. [Morgulis, p. xvi.]

Thermodynamically directed chemical evolution could conceivably proceed indefinitely without changing from a non-living to a living state. Only when organic matter had achieved a high degree of organization, and had acquired diverse propensities through the concatination of such substances (with chance as the only arbiter) did primordial life emerge as a new dimension in nature: matter perpetuating its own organization. Natural selection, operating upon chance variations, set the evolutionary direction along numerous pathways which living things have followed irresistibly. [Morgulis, p. xxii.]

From Lifeless to the Living

If we ignore the rigorous standards of physics for the moment, we can argue that this universe is multidimensional—not simply a space-time scheme. A possible dimension that might require additional natural laws is Consciousness; another is Life. The latter certainly involves biochemical regulations more complex than are obvious in the laws of the standard space-time world. Oparin points to the new properties that have been uncovered in biochemical evolution, the new colloidal-chemical order that has been imposed on the more simple organic chemical relations. These new properties are a cue, a clue, and a challenge. They may lead to a description and definition of life that satisfies chemist, biologist, and perhaps even philosopher.

This brief survey purports to show the gradual evolution of organic substances and the manner by which ever newer properties, subject to laws of a higher order, were superimposed step by step upon the erstwhile simple and elementary properties of matter. At first there were the simple solutions of organic substances, whose behavior was governed by the properties of their component atoms and the arrangement of those atoms in the molecular structure. But gradually, as a result of growth and increased complexity of the molecules, new properties have come into being and a new colloidal-chemical order was imposed upon the more simple organic chemical relations. These newer properties were determined by the spatial arrangement and mutual relationship of the molecules. Even this configuration of organic matter was still insufficient to give rise to primary living things. For this, the colloidal systems in the process of their evolution had to acquire properties of a still higher order, which would permit the attainment of the next and more advanced phase in the organization of matter. In this process biological orderliness already comes into prominence. Competitive speed of growth, struggle for existence and, finally, natural selection determined such a form of material organization which is characteristic of living things of the present time. [Oparin, pp. 250-251.]

The origination of life was a transition from organic to biological chemistry, from lifeless to living matter, from the inanimate to the animate realm of Nature. But what is Life? Is it some new property of organic matter acquired in the course of evolution or is it something which resulted from the organization of organic matter? Irritability, motility, growth, reproduction may be good aids to differentiate a live from a dead organism but it is questionable whether these represent the fundamental properties of primordial life. There is good reason to think that a certain period of the Earth's history must have been marked by complete sterility, i.e., absence of organisms; therefore, the fundamental property or properties of living systems must have appeared in highly complex protein macromolecules antedating the appearance of cellular organisms.

Proteins containing nucleic acid are the only constituents of organisms which are known to possess the capacity to grow and to reproduce directly by self-duplication or by replication. But as organic compounds they can neither grow nor reproduce. Neither viruses nor genes, both of which represent nucleo-protein systems, can duplicate or replicate themselves unless they are incorporated within a suitable cell or nucleus. Considered simply from the point of view of capacity to reproduce, are these nucleoproteins living or non-living systems? [Morgulis, p. xi.]

A touch of Darwinian evolutionary principle has entered the primitive precellular phase of life. The survival of the best adapted prevails not only with plant, man, and beast, but also in the microcosmos.

Natural selection has long ago destroyed and completely wiped off the face of the Earth all the intermediate forms of organization of primary colloidal systems and of the simplest living things and, wherever the external conditions are favorable to the evolution of life, we find countless numbers of fully developed highly organized living things. If organic matter would appear at the present time it could not evolve for very long because it would be quickly consumed and destroyed by the innumerable microorganisms inhabiting the earth, water and air. For this reason, the process of evolution of organic substance, the process of formation of life sketched in the preceding pages, cannot be observed directly now. The tremendously long intervals of time separating the single steps in this process make it impossible to reproduce under available laboratory conditions the process as it occurred in nature. [Oparin, p. 251.]

The conditions of the Earth during the past couple of billion years have undergone such radical alterations that biogenesis may no longer be possible. However, as Oparin points out, even if biogenesis were operating at the present time, the innumerable predatory organisms which populate the Earth would quickly destroy the products of biogenesis. [Morgulis, p. x.]

The concluding paragraph of Oparin's volume carries a note of caution about celebrating too soon, and a note of conviction that the goal is attainable.

We are faced with a colossal problem of investigating each separate stage of the evolutionary process as it was sketched here. We must delve into the properties of proteins, we must learn the structure of colloidal organic systems, of enzymes, of protoplasmic organization, etc. The road ahead is hard and long but without doubt it leads to the ultimate knowledge of the nature of life. The artificial building or synthesis of living things is very remote, but not an unattainable goal along this road. [Oparin, p. 252.]

Theory Put to the Test of Experiment

Because of the historical importance of Stanley Miller's work, mentioned above, I give in full the abstract of his report.[*] The highly technical details and specialized language will indicate to the reader the sophistication and difficulty of biochemical research on problems dealing with the emergence of living organisms.

A mixture of gases, CH_4, NH_3, H_2O and H_2, which possibly made up the atmosphere of the Earth in its early stages, has been subjected to spark and silent discharges for times of the order of a week to determine which organic compounds would be synthesized. Several designs of apparatus and reasons for their construction are described. Analyses of the remaining gases were made and CO, CO_2, N_2 and the initial gases were found. A red compound that seems to be associated with the trace metals is formed, as well as yellow compounds, probably polymers, which have acidic, basic and ampholytic properties. The mixture of compounds is separated into acidic, basic and ampholytic fractions with ion exchange resins. The amino acids are chromatographed on Dowex-50 and the acids on silica. Glycine, d, l-alanine, β-alanine, sarcosine, d, l-α-amino-n-butyric acid and α-amino-isobutyric acid have been identified by paper chromatography and by melting points of derivatives. Substantial quantities of several unidentified amino acids and small amounts of about 25 amino acids are produced, while glycolic, d, l-lactic, formic, acetic and propionic acids make up most of the acid fraction. Quantitative estimates of these compounds are given. Evidence is presented that polyhydroxy compounds of unknown composition are present. HCN and aldehydes are direct products of the discharge. Although there is insufficient evidence, the synthesis of the hydroxy and amino acids may be through the hydroxy and amino nitriles in the solution. The relation of these experiments to the formation of the Earth and the origin of life is briefly discussed.

From his cautious discussion of the results of the experiments, which indicate what might have happened a few thousand million years ago, I excerpt a few sentences.

If these experiments are to any degree a representation of the reducing atmosphere of the Earth, then we see that not only would the formation of organic compounds be easy, but that most of the carbon on the surface of the Earth would have been in the form of organic compounds dissolved in the oceans. . . .

These ideas are of course speculation, for we do not know that the Earth had a reducing atmosphere when it was formed. Most of the geological record has been altered in the four to five billion years since then, so that no direct evidence has yet been found. However, the experimental results reported here lend support to the argument that the Earth had a reducing atmosphere; for if it can be shown that the organic compounds that make up living systems cannot be synthesized in an oxidizing atmosphere, and if it can be shown that these organic compounds can be synthesized in a reducing atmosphere, then one conclusion is that the Earth had a reducing atmosphere in its early stages, and that life arose from the sea of organic compounds formed while the Earth had this atmosphere. This argument makes the assumption that for life to arise, there must be present first a large number of organic compounds similar to those that would make up the first organism. . . .

This reasoning, coupled with the independent argument that hydrogen is so abundant in the universe, places the assumption of a reducing atmosphere on sufficiently firm basis that it should be taken into account in future discussions of both the formation of the Earth and the origin of life.

A highly important extension of Miller's work on the synthesis of the organic compounds that must have been the forerunners of living organisms is reported from the Geophysical Laboratory of the Carnegie Institution of Washington by Philip H. Abelson. In setting up hypothetical "primitive atmospheres," he tried out mixtures other than the methane, ammonia, water vapor, and hydrogen used by Miller, with the result that always amino acids were synthesized. For example, he successfully replaced ammonia with nitrogen and methane with carbon monoxide and carbon dioxide, gases that the early volcanoes must have supplied in abundance. Here is the compact abstract of his report (*Science*, Nov. 9, 1956):

Simulating atmospheric conditions that might have been present early in the history of the earth, amino acids such as alanine, β-alanine, glycine, and sarcocine have been synthesized employing a variety of compositions. Combinations of gases, including $CO_2 - N_2 - H_2 - H_2O$, $CO - N_2 - H_2 - H_2O$, $CO_2 - NH_3 - H_2 - H_2O$, were subjected to electric discharges, and in each case amino acids were formed. The earlier work of S. Miller employing $CH_4 - NH_3 - H_2O$ has been confirmed.

Several other recent contributions from the physiological, chemical, and virological laboratories could be cited, as well as earlier shrewd speculations on the origin of life, but the foregoing quotations and arguments should suffice to show that biogenesis is no longer a hazy field of investigation. The evolution of macromolecules is natural, and apparently eager! As Dr. George Wald has remarked, just give the right molecules a chance in suitable environment; we do not have to do everything for them; they do a great deal for themselves.

★ 10 ★

What Should Be the Human Response?

The cosmic drama that is unfolding in the chemical laboratories, and through the astro-geo-bio-physical researches in field and observatory, incites renewed curiosity about the "neurotic equipment" of this researching being—curiosity about the human mind, how it guides and misleads, and how it compares with other animalian minds. In this concluding chapter we shall refer briefly to the sense receptors in the animal world. Also we shall consider some of the grim future hazards of the human race, and ask what should be the rational response of man to his current acquaintance with these matters. The answer should properly come from the professional philosopher, theologian, and humanist, but also from the thoughtful layman. It is his response that most concerns us.

Again this writer should state that the task of the proper scientist is to bring forth and explain as best he can the raw materials from which other analyzers can fabricate philosophies and suggest, if they will, programs and goals.

Noting again the subtitle of this volume, let us see if we can itemize some human responses to the cosmic facts that have emerged in recent times, or at least indicate typical responses.

One's first reaction to knowledge such as here reviewed concerning stars, space, time, and man is likely to be amazement and incredulity, followed by a desire to hear no more about such things. "They are too confusing and we are busy with life."

Such a response appears to be common, but fortunately it is often temporary, and is generally followed by curiosity, surprise, receding doubts, and then by deep respect and wonderment. In this second stage questions are asked by those who want to hear and think beyond the simplest facts.

"What does this mean about my own importance in the world?"

"What does it mean with respect to the prevalent religious teachings, whether the standardized, the evangelical, or the softly routine teaching?"

"Why should we make this adjustment to science (which appears to have some fixations of its own) when a retreat from the turbulent front into superstition or indifference still leaves us with our exciting lives and some comforting myths?"

The third stage in the response to the voice of the universe is, I hope and believe, completely rational. It is as intelligent as our mediocre equipment of sense organs and our mental capacities allow. We begin to think brave thoughts, and ask hard questions, seeking at least partial answers, such as those offered here.

Growth Through Understanding

To repeat an earlier conclusion: We are incurably peripheral. We are primitive in a sensory sense. With help from our Star we have slowly evolved from the wonder-working Archaeozoic ooze in which so many biological experiments have been made. We have risen from the primeval "Hot Thin Soup," from which also evolved bluebirds and roses, and a million other to us less pleasing but wonderfully constructed organisms. We must henceforth live with awareness of these cosmic facts and of our ancestry, no matter how disturbing such knowledge is to rigid creeds. With much less convincing evidence than now at hand, we have been vaguely aware of our immediate anthropoid ancestry for nearly a century. We have been spoken to occasionally about the truths of stars and life, but mostly we do not listen closely, or act.

The cosmic immensities, whether of space and time, or of outlook and concept, should, however, not dismay us, the local gropers and interpreters. In our natural program of growth through understanding, each day competes with our yesterdays. Fortunately for us that competition, that striving and groping, is largely inborn, nicely automatic; our succeeding days compete as a matter of course. If care is taken to cope with the natural regressions that often ensue from static conformity, we shall continue to evolve with the rotating of the planets and the radiating of the suns. We grow naturally with the passage of time, as do the animals and plants. They also make the effort, through adjustments and improvements, to survive and grow. The tempo of their evolution, however, is often even slower than ours, and our hominid evolution appears to have been a very tedious travel through the Cenozoic Era.

But this automatic, slow, slight, and hesitant rising is now not enough for us—the considerably Intelligent and somewhat Informed. We can consciously speed the development. It is not growth in size, or strength, or longevity, but growth primarily in the qualities that we associate with mind, a development that includes those fine indefinables—heart and spirit. And therein lies the nucleus of our cosmic ethic. The evidence clearly shows that we have the potentiality not only of conforming to the cosmic theme of Growth, but we can perhaps elaborate or revise some of the natural rules. Indeed, each day can and should compete with all the yesterdays of the species.

The Generic Mind

We who read, write, and contemplate have minds that have been laboriously taught. The teaching has been done by books and teachers, and by our own efforts. Through such instruction we have attained a degree of competence. We can get around physically with reasonable safety, and react to our environment intelligently and with some pride. But without the instruction that started a few hours after birth we would not be doing very well. We have needed help from the beginning—in decreasing amount, to be sure, but nevertheless we have been dependents all our lives. Instincts, such as we attribute to the newly hatched, uncared-for mosquito or housefly, were not, at our beginning, available to keep us alive the first week, even if we had been physically competent. When neglected, we could only howl instinctively for nourishment. A very primitive performance. Later a more subtle "howling" for food was taught us. That subtlety was a part and product of our training. The housefly was born with her training completed. She uses not only her own nervous ganglia for the planning and executing of life's operations, making emergency decisions and acting thereon, but she uses also and mainly her generic mind—something that we largely lack.

Probably more of our own behavior is instinctive than we know of, and perhaps more than we would care to admit, being so proud of the capacities of our educated minds. The ratio of noninstinctive to instinctive action, however, is high for us, low for the fly. We can put it another way—the ratio of our apparently personal decisions to the decisions that the experience of the race has built into us is high because so little is built in. The housefly makes some decisions of her own, but mostly uses the gradually acquired, the slowly trained mind of thousands of generations of flies.

In view of the width of the cosmos and the slim hold on existence that Nature has provided for Homo sapiens, it would seem properly modest if we talked less about man being superior, less about his being the anointed of the gods. Who was anointed, we ask, throughout the half billion manless Paleozoic and Mesozoic years when thousands of kinds of wonderful animals sought survival on the Earth?

Our claimants for human distinction often say that man's superiority arises from his "historical sense." They have in mind, probably, the histories written by such as Gibbon, Parkman, and Toynbee. Or, thinking a little deeper, they also have in mind the word-of-mouth transmission of the unwritten folklore of the past few thousand years. And there is unwritten history still more basic in the mother's murmuring of do's and dont's to her dimly responding infant; she is setting up his "historical sense." But how does this differ, except in degree, from the sparrows chirping to their young, or the worker ants twiddling antennae with the newly emerged callows? And is not the impelling sex urge of a million kinds of animals one phase of the reciting of the deep and moving history of the ages—reciting it more profoundly, more insistently than it can be done through symbols scratched by the higher and vainer primates on cavern walls, or by the binding of their more systematic scratchings into octavo volumes?

Terrestrial man's intellectual responses differ only in degree from those of other earthly organisms; compared with the best responses by the "highest" extraterrestrial sentient beings, they might incite in us small pride.

"But only man," you may argue, "*acts* on the basis of his historical knowledge." Again nonsense. In the first place he does not do it very well—goes right on fighting thoughtless futile wars, goes on acting more beastly than angelic. History appears to teach him little. And secondly, after their fashion, most animals also act on the basis of experience.

The survival of the so-called fittest is a response to an organism's historical knowledge.

There can be no better laboratory for the elaboration of thoughts on man's orientation in a complex world than a flowering meadow, or a noisy brook, or a spiral galaxy. For the green leaves are sucklings of a star's radiation. The rapids in a brook, responding to universal gravitation, perform erosions of a sort that have worn down to oblivion the lofty pre-Alps and the primitive Appalachians. The hundred-ton maple tree that calmly dreams through the decades is in the same universe as the Andromeda Galaxy with its billions of seething stars. The tree heeds the impulse of gravity according to the same rules as those subscribed to by the stars in a globular cluster. Further, it is made of the same complex molecular aggregates as are the birds in its branches, the parasites at its roots, and the scientists who wonder about it.

In a complex situation one simple fact stands out: we must link ourselves with all the others that participate in life; we must go beyond life and associate ourselves continually and insistently with the solid rocks of the earth, the gaseous winds of the sky.

Of course, it is our privilege to fancy ourselves as the thinkers and prognosticators for all earthly organisms of the past, present, and future, for all the stars and nebulae, for all the basic entities. It could be one of our hallucinations that we are dominant because we can think and make a pattern for all the world.*

---

* May it not be true, however, that to the tent caterpillar God is gray—the Great Devourer, the Eternal Eater? William Butler Yeats reports that apparently the moor fowl, the lotus flower, and the roebuck all claim that God is made in their respective forms, and adds that he

"passed a little further on and heard a peacock say:
Who made the grass and made the worms and made my feathers gay,
He is a monstrous peacock, and he waveth all the night
His languid tail above us, lit with myriad spots of light."

The close student of social insects would not boast about the superiority of man's social awareness; and he might even qualify the claim of superiority for the human brain. He has seen too much of the wonderful—this student of animal societies. He has seen the honey bee dance her complex geometry, instructing by sight and scent and diagram her student gatherers of honey and pollen. He has witnessed the magic of many tiny insects carrying out complicated enterprises. Even a casual observer would be less boastful of human superiority if he did not have habitual recourse to an easy cliché: "It's their blind instinct. They are thoughtless. They cannot reason. They don't even have a proper brain."

A little analysis cripples that routine assertion, for the "blindness" more properly refers to the observer (and he could better say glorious instinct instead of blind). The "thoughtlessness" is an assumption, groundless and erroneous—or it implies a stricture on the meaning of "think." And "proper brain" is a rather careless acceptance of the hypothesis that only a neural assembly that is contained in a skull is "proper."

"We alone can reason," is one of the other totally unreasonable assumptions. What evidence is there of thoughtlessness and unreason in the bird choosing its nesting site or a spider locating her web? Their generic minds do much, and thoughtful adjustment to immediate situations does the rest. In contrast, we higher primates are short on inheritance but long on ability to cope with the environmental unusual. A matter of degree or intensity is here implicated in not com-

---

He might have added:

"At last I reached the darkest spot within the wooded glen;
 From out the murk a voice bespoke the faith and boast of men:
 The God that drives the world around, makes stars that light the sky,
 He's in man's proper image formed—his greatest work is I!"

And, seriously, Mary Baker Eddy asks: "What is the God of a mortal but a mortal magnified?"

pletely different kinds of thinking and reasoning. In certain characteristics and skills we do not excel; we only approach the abilities of other animals, and sometimes not very closely; and in other characteristics and skills, such as reading books, they approach us, but not closely. In differing degrees the higher mammals have all of our own virtues and vices, our abilities and futilities.

The teaching of all this is: Don't take man too seriously, even when orienting him among the animals and plants on this local planet; and certainly not when comparing him with possibilities elsewhere in the richly endowed Metagalaxy.

But let us not tire ourselves with annoyance at man's egocentric vanities; and rather than abuse the presumptuous primate, I should simply call attention to the existence of the generic mind, the most precious inheritance of most animal forms. I point also to the success, over the millions of years, of thousands of animal species that have acquired security by methods that are lost to us, lost to animals that must resort, for survival, to mother murmurings, folk tales, and printed history.

The foregoing rough evaluation of those who write and read is offered only as one item in the placement of mankind in the biological sequences and possibilities. Man is involved in the higher biochemistries, and in complicated neurotic reactions, as well as in time and space. Orientation has many facets.

On Our Limited Sense Organs

In earlier chapters we have more than once intimated that sentient organisms, the product of biochemical evolution, must be a common occurrence in the universe. From general considerations of planetary origins and the evolution of chemical compounds on a cooling planet's surface, we concluded first that not less than a hundred million "high life" locations exist, and that the number is probably more like a hundred trillion. Secondly, that there is no reason not to believe that the biochemical evolution on, let us say, one half of the suitable planets has equaled or attained much greater development than here. Thus, in answer to an earlier question, we have decided that *we are not alone* in this universe, and we have by implication suggested that the Omnipotence (shall we say Nature?) which looks after us has very much else to do. But before we turn to such contemplations, I should like to comment on the tools of comprehension. Again we may find that our self esteem is healthily eroded.

Naturally we wonder if it is not likely that some of those distant sentient organisms, which bask and grope in the light and warmth of well-placed stars, may not have succeeded better than we in the search for a world dynamic. We have not really gone very far—blocked on all sides as we are by the yet unknown, perhaps by the unknowable. Most of us know only what we see in print. Or what we hear someone say. The senses of seeing and hearing provide our best methods of ascertaining what is what, and why. The eyes and the ears—without them it would be a strange world. With better eyes and ears, * and with additional sense organs, we might have attained a much finer knowledge than we have up to now.

* But if our ears were exceedingly sensitive to all wave lengths, and the physics of the medium permitted, might we not then hear the molecules of the air banging into each other? And what a din! Frightful static! It could easily overwhelm all other sound. We would hear no music, no shouting, no speeches! Much as fog and smog limit the effectiveness of the eye, the clatter of the molecules would limit the effectiveness of the ears.

The most important aid to the human mind in the understanding of the universe, both the microcosmos and the macrocosmos, is the electromagnetic spectrum, as currently exploited. The major part of our knowledge of the universe has in the past come through information provided by one sense organ alone—that of vision. Our eyes are, however, sensitive only in a small segment of the long radiation spectrum—they are sensitive from the violet to the red—much less than two octaves. But with a suddenness unequaled in the development by artifice of sense-organ spread, we have now learned to explore Nature with radiations extending over a range of more than fifty octaves, a range from the cosmic rays of less than a billionth of an inch in wave length, through gamma rays, X rays, and the ultraviolet, up to the blue-to-red radiation our eyes record; and then from red to heat waves to radio and on to electric wave lengths measured in miles. We know and measure and use these off-color radiations not directly with the retinas of our eyes, as we do with light, but with artifacts, with the retinas, we might say, of photographic plates, Geiger counters, and photo cells.

The eyes and other sense organs arose naturally to serve animals in the practical problems of existence, not for use in profound researches into the nature and operations of the universe. Practical existence had not until recently included the thirst for "impractical" knowledge. Our intellectual desires have gone ahead of our built-in sensory receptors.

It happens that the range of human vision from violet to red covers that part of the radiation spectrum where the sun's light is most intense. In fact, there is not much solar intensity in the short waves of X rays or the long waves of radio.

If there are animals with vision on a planet near a hotter, bluer star than the sun, for instance Rigel in Orion, they probably are more sensitive than we are to light in the bluer section of the spectrum; and near a cooler, redder star, like Betelgeuse, more sensitive to reddish light. Of course, our sun was not made yellowish just to suit our visual sensitivity! On the contrary, our vision has evolved to fit our star's most abundant radiation.

A fact worth emphasizing at this point is that man has no built-in sense organs for either long wave lengths or for those short-wave radiations that are more violet than the violet. He is blind, except to a narrow range in the electromagnetic spectrum. If he had been outfitted from the beginning with sensitive recorders throughout all wave lengths from hard X rays to long-wave radio, his knowledge of the world might have differed vastly from that which he has slowly developed by way of his restricted vision. Earth waves, dark lightning, molecular motions—such phenomena might have long been in his commonplace book if he had been adequately equipped. To feed himself, to dodge or overcome his enemies, and to find his mate, he has had little need of the solar radiation with wave lengths like those that activate our radios; and since the atmosphere, as it has developed on the earth, has through its ozone blockage shut off the ultraviolet, he, as an animal, has also had no practical need for the short-wave end of the spectrum. It is only as the dominant interpreter in the earth's present Psychozoic Era that he has developed the new sensory needs and devised the artifacts to meet some of them.

Even when he supplements the good sense of vision with a sense of hearing, a poor sense of smell, and a complex of tactile senses, man still is not too well equipped to cope with all the cosmic mysteries. In fact, as an organism ambitious to know, and know deeply, he is rather primitive in his senses, if not in sense. (Our primitivism in body anatomy is, of course, generally recognized.)

Our sense organs, I should for emphasis repeat, are limited in number, in range, in effectiveness. Every human sense receptor, except possibly that concerned in tone discrimination, is outdone by the corresponding receptor of one animal or another—by the hawk's vision, the dog's hearing, the insect's smelling. Some stars have enormously strong magnetic fields; ours a weak one, and we have no recognized sense receptor whatever for magnetism.

But these sensory limitations, and the resulting failure to comprehend fully much of Nature, may be only a local deficiency. On the basis of the new estimates of the great abundance of stars and the high probability of millions of planets with highly developed life, we are made aware—embarrassingly aware—that we may be intellectual minims in the life of the universe. I could develop further this uncomfortable idea by pointing out that sense receptors, in quality quite unknown to us and in fact hardly imaginable, which record phenomena of which we are totally ignorant, may easily exist among the higher sentient organisms of other planets.

Sometimes we suspect the existence of senses other than those we recognize in ourselves, among animal and plant forms on this planet—not merely extended ranges of hearing or of vision or smell, but entirely different responses. The bees and ants respond, as we do not, to polarized light; the birds in migration—to what? And there are those among us who dream of vestigial or embryonic senses hovering about the human psyche. *

Humility Comes Naturally

Rather than dwell on these probabilities, let us note simply that anthropocentric religions and philosophies, which have so often been conspicuously earth-bound and much tangled up with the human mind and human behavior, have in these

---

* Animal sense organs are ably discussed by J. D. Carty in *Animal Navigation* (London, 1956).

present days an opportunity for aggrandizement through incorporating a sensibility of the newly revealed cosmos. If the theologian finds it difficult to take seriously our insistence that the god of humanity is the god of gravitation and the god of hydrogen atoms, at least he may be willing to consider the reasonableness of extending to the higher sentient beings that have evolved elsewhere among the myriads of galaxies the same intellectual or spiritual rating he gives to us. A one-planet deity has for me little appeal.

You may say that these are but speculations, insecurely founded, and that you choose to believe and reason and worship otherwise. And I must reply that you should follow your inclination. But you are invited to think seriously of the cosmic facts. Let us hope that an easement is not sought in comforting tradition alone, or in resort to crude irrationality. The new knowledge from many sources—from the test tube, from the extended radiation spectrum, the electron microscope, experimental agriculture, and the radio telescope; from mathematical equations and the cosmotrons—the revelations from all these make obsolete many of the earlier world views. The new discoveries and developments contribute to the unfolding of a magnificent universe; to be a participant is in itself a glory. With our confreres on distant planets; with our fellow animals and plants of land, air, and sea; with the rocks and waters of all planetary crusts, and the photons and atoms that make up the stars—with all these we are associated in an existence and an evolution that inspires respect and deep reverence. We cannot escape humility. And as groping philosophers and scientists we are thankful for the mysteries that still lie beyond our grasp.

There are those who would call this attitude their philosophy, their religion. They would be loath, I hope, to retreat from the galaxies to the earth; unwilling to come out of the cosmic depths and durations to concern themselves only with one organic form on the crust of one small planet, near a commonplace star, at the edge of one of the galaxies. They would hesitate to retreat to that one isolated spot in their search for the Ultimate. May their kind increase and prosper!

Fish or Homo—That Is the Question

The protoplasmic experiment, as we may thus casually refer to life, so far as it has progressed on this planet is an inspiring demonstration of Nature's intricacies. We are probably much more complex in body and mind, much more wonderfully made than we know. The human imagination, versatile as it is, would be hard put to dream up such complexities and coordinations as those achieved inside the simplest living cell, and it would be equally far surpassed by full knowledge of the true mechanism in the center of a single molecule. Fiction lags far behind the facts. The yet undiscovered, the still unknown but not unknowable, so far "transcends the what we know" that a very happy future of inquiry and discovery lies ahead.

Limited as we are in both wisdom and information, we are nonetheless emboldened by our attainment of a rough coarse understanding of this world that presently exists and long has been; we are encouraged to imagine and forecast the world that is to be.

Man's concern with his own future as an individual is instinctive and it is often intense. It could well be lessened. The actuarial tables tell him the number of his days. His concern for the future of the *species* is now slight. It could well be increased. To me it is a sign that we are sincere subscribers to the "growth" motif, to the growth of mind if not of body, to the evolution of the race if not of the individual, when we inquire what lies ahead for mankind— what lies far, far ahead in the times when the galaxies will have scattered, the moon receded into its coming faintness, and the massive mountains will have been worn away by the persistently weathering winds and rains.

Since men in the future must continue to rely on their wits and acquired wisdom, and cannot depend on a built-in generic mind, and since they are embattled in a continual contest with a Nature that includes their dangerous selves, we are led to worry and wonder whether man or the meek will eventually inherit the earth. As an example of the meek I might choose the fish. They chiefly employ instincts and not heavy forebrains.

Fish or Homo, that is the question. Which animal type will most certainly be here 10,000 years from now, and which will more likely fall the victim of fate and folly? The answer of course is too obvious. The fish have been here several hundred million years; man but a few hundred thousand. The oceans are stable enough in their salinity, temperature, and food supplies to suit indefinitely a thousand species of fish. It it difficult to imagine a way of curtailing the life of that class of animals without complete disruption of the planet, or the poisoning of the plankton food of all the seven seas. But 10,-000 years is a pretty long time for Homo. His structure and social manners do not make him a good insurance risk.

What agency, we ask, might terminate the human race? Let us survey the possibilities.

The Opposition to Survival

Nearly three-fourths of the earth's crust is under the oceans; the remainder protrudes above the water level to various extents. There is some falling and rising of the shore lines. The mountains are worn down by the winds and rains, and rise up through the wrinkling of the earth's crust. In general, however, the continents seem to be pretty stable over the geological eras. Extrapolating into the future, we should say that it is extremely unlikely that man will be drowned out by the sea or desiccated by the lack of water. A man-eliminating deluge is not in the cards, nor is a totally lethal drought.

Let us look farther into the problem of completely exterminating the human race, taking the look as an exercise both in scientific objectivity and in speculation. Let us see in what manner we, as agents of Nature (or the devil!), could proceed to the elimination of Homo sapiens. First we shall set a time limit—not too short, not too long. It appears very likely that the human race will still be here a century hence, and not very likely here in a hundred million years; probably it will still be a going race a thousand years from now, but not surely for a hundred thousand.

We shall choose to examine the probability of men being on the earth's surface ten thousand years from now. Most of the following considerations have a scientific validity and small fancy; others are necessarily large with imagination and, as yet, small in scientific backing.

We begin with the macrocosmic instruments of eradication. What is the possibility of the collision of the earth with stars? Stars exhibit a considerable degree of randomicity in the directions of their motions and move with speeds averaging in our neighborhood some twenty miles a second. If the earth or even the sun were struck by one of these bodies our project would be completed; terrestrial biology would be finished. But the stars are so widely separated that collisions are out of the question in our chosen relatively short time interval of ten thousand years. The probabilities are overwhelmingly against trouble with stars.

We appreciate that escape, and ask about the sun cooling down enough to freeze us out, or blowing up into a nova and incinerating the planets. No likelihood at all, or at least highly improbable, for the sun appears to be of the relatively quiet stable type of star, and its radiation has been steady for many hundreds of millions of years. Its hydrogen content is ample to supply radiant energy, produced by atomic fusion, for a million times the duration we require.

Safe from annihilation by stars and sun, should we fear the earth's misbehavior, such as its abandoning orbital regularity, getting too near the sun or too far away? The answer is no. Our mathematical analyses show that the planetary orbits are completely stable over time intervals such as we are here considering. The earth moves in what is practically a vacuum in a nearly circular path around the sun, and neither will its daily rotation nor its yearly revolution change perceptibly in the allotted hundred centuries. (We could of course adapt ourselves to small changes, as we were able to adapt ourselves to the coming and going of the ice sheets that occurred in the northern hemisphere during the past hundred thousand years.)

Already we have mentioned the relative constancy of the continents and oceans. Terrestrial life has readily adjusted itself to the ups and downs of land and sea in the past million years, and in our ten thousand years the slow moving mountains and shore lines will present little hazard. Man can certainly keep ahead of the moving ice and the spreading deserts. And of course he has easily survived volcanism, storms, and floods, and will continue to do so if he remains moderately intelligent.

To poison the atmosphere with overabundance of volcanic gases and make it unbreathable by land animals, including man—well, such has not happened in the past five hundred million years and it is certainly unlikely in the next ten thousand; the earth is gradually getting over its eruptive birth pangs.

The poisoning of the atmosphere by interstellar gas and dust is a very long chance for two reasons; the interstellar gas is mostly nonpoisonous hydrogen and helium, and its abundance is so little that our own atmosphere shields us from the gas, as it now also protects us from the tiny high-speed interplanetary meteors.

To summarize the progress so far, in the project of eliminating man (and other animals) from the earth's surface, we get no likely help from the stars, from interstellar dust, from the sun's radiation or its lack, from the deviation of the earth from its present orbit, from deadly climates, or from the chemistry of the earth's land, air, and water.

We turn to the biological sciences. The large beasts are no longer a threat, nor any of the plant and animal forms. We are now competent to cope with bacteria, viruses, and the like, at least sufficiently well to keep our species going. Of course, something fatal and world-wide could happen, the disaster coming to us anywhere from star crash to infective protein; but the chances are heavily against it—less than one chance in a million for trouble with astronomical bodies; less than one in a thousand for serious difficulties with climates, volcanoes, world-wide floods, or desiccations; and perhaps less than one in a hundred for planet-wide incurable disease.

(Even if ninety-nine per cent of the world's population should fall foul of pandemic disaster, there would still be left more than twenty millions, reseeding the earth with humans; the total elimination effort would have been a failure. Spoiling a civilization is one thing, and perhaps not too arduous; complete eradication is quite another, and vastly more difficult.)

In other words man seems to have a healthy prospect, a long security from stars, climate, and terminating germs. But wait! I have not named the real danger, and it is bleakly ominous, as everyone in these days agrees. The danger is man himself. He is his own worst enemy. He is acquiring tools and studying techniques that might solve the problem we set—the complete elimination of Homo from the planet earth.

Much could be written on various methods for man's elimination of himself, and something ventured on the defenses against this grim danger. But that is not a responsibility of this particular essay. Rather, it is the responsibility of everyone who desires to justify, for our species of Homo, the name sapiens.

Confessing to Optimism

To me it seems proper to conclude this essay on a note of humility and hope, if not of high confidence. Certainly we should be humble about our trivial accomplishment in understanding the total of the external world. We know enough to get along, as do most of the other animals. We can cope with all of the primitive challenges. And going farther we can construct new worlds of ideas and beauty. Our number and our works are impressive, although both number and works are limited to the surface or near the surface of one planet.

I assume that the human mind and heart will successfully confront the hazards of mankind as they arise. This habitation on a pretty steady planet is comfortable on the average, and may get happier. We have increased the length of our useful lives. We have built up ethical systems that average to bring us safety and satisfaction, but which greatly alarm and dismay us by frequent failures. We know that the rules of the stars are hard, that the flow of time is irrevocable, that death is dark and will accept no substitutes. But even so, the lights can, if we cooperate, exceed the shadows. The imagination can enter when knowledge falters. We of the higher primates have delved into the half-known cosmic facts deeply enough to recognize also the need of cosmic fancies when facts are delayed.

It is my own belief (or is it fancy?) that the central theme of biological existence is to grow in refined complexity, in durability, in adaptability. Man as half beast, half angel must comply with the biogenetic common law, but he is able to make amendments thereto.

As rational practitioners of life and tentative interpreters of the cosmos, we deplore superstition—the last stronghold of the irrational. But, thanks to man's reasoning, never before has hampering superstition been in retreat on so wide a front. Belief in the supernatural is tempered with thought. Rationalism has captured many outposts in our necessarily continuous conflict with the Tyranny of the Unknown. We no longer need appeal to anything beyond nature when we are confronted by such problems as the origin of life, or the binding forces of nucleons, or the orbits in a star cluster, or the electrochemical dynamics of a thought, or some super-entity of the material universe. We can assail all such questions rationally.

In scores of ways improvements are possible in the three principal fields of human activity—physical, mental, social. In these fields we have accomplished many things since the Pliocene; we can do much more. It is probable that the men of the future will overcome our shortcomings and build out of our thoughts and acts a finer mental and social structure—one that is in better keeping with Nature's heavy investment in the locally dominant human race.

IF YOU HAVE CHILDREN, they will be required to read certain classics as part of their school-work. Now, at last, you can have these classics in handsome, inexpensive editions for your home library. Get them now and be prepared. For example:

————FOR ONLY 35c EACH————

[more→]

How to Build
A Low-Cost Library

You can build a personal library of the best books for as little as 25 or 35 cents a volume. Choose from thousands of the classics and best sellers in literature, biography, poetry, art, history, religion, reference and science as listed in a new catalog:

Paperbound Books in Print

If you've often had trouble finding the paperbacks you want, here are over 6,000—with information on how and where to get them. Here you can locate all the low-priced paper books available either by checking the thousands of titles listed alphabetically by author and by title, or by looking under any of the 60 categories where selected titles are grouped under helpful subject classifications.

Order your copy of this unique buying guide today—either from your dealer or direct from Mail Service Department, Pocket Books, Inc., 630 Fifth Avenue, New York 20, N.Y.

Make checks payable to: R. R. Bowker Company. Single copies are $2 net prepaid, or you may get two semi-annual issues for $3.